Phytochemistry of Plants of Genus *Piper*

Phytochemical Investigations of Medicinal Plants

Series Editor:
Brijesh Kumar

Phytochemistry of Plants of Genus *Phyllanthus*
Brijesh Kumar, Sunil Kumar and K. P. Madhusudanan

Phytochemistry of Plants of Genus *Ocimum*
Brijesh Kumar, Vikas Bajpai, Surabhi Tiwari and Renu Pandey

Phytochemistry of Plants of Genus *Piper*
Brijesh Kumar, Surabhi Tiwari, Vikas Bajpai and Bikarma Singh

Phytochemistry of *Tinospora cordifolia*
Brijesh Kumar, Vikas Bajpai and Nikhil Kumar

Phytochemistry of Plants of Genus *Rauvolfia*
Brijesh Kumar, Sunil Kumar, Vikas Bajpai and K. P. Madhusudanan

Phytochemistry of *Piper betle* Landraces
Vikas Bajpai, Nikhil Kumar and Brijesh Kumar

For more information about this series, please visit: https://www.crcpress.com/
Phytochemical-Investigations-of-Medicinal-Plants/book-series/PHYTO

Phytochemistry of Plants of Genus *Piper*

Brijesh Kumar, Surabhi Tiwari, Vikas Bajpai
and Bikarma Singh

CRC Press
Taylor & Francis Group
Boca Raton London New York

CRC Press is an imprint of the
Taylor & Francis Group, an **informa** business

First edition published 2020
by CRC Press
6000 Broken Sound Parkway NW, Suite 300, Boca Raton, FL 33487-2742

and by CRC Press
2 Park Square, Milton Park, Abingdon, Oxon, OX14 4RN

© 2020 Taylor & Francis Group, LLC

CRC Press is an imprint of Taylor & Francis Group, LLC

ISBN: 978-0-367-85757-8 (hbk)
ISBN: 978-0-367-50056-6 (pbk)
ISBN: 978-1-003-01487-4 (ebk)

Typeset in Times
by codeMantra

Contents

Preface

Plants are considered as one of the most important sources of modern medicine. They have been used for different ailments of human beings worldwide since the beginning of civilization. Bioactive secondary metabolites have been considered as a fundamental source of medicine for the treatment of a range of diseases in modern medical system. India has a rich heritage of medicinal plants which are the basis of indigenous systems of medicine such as Ayurveda, Siddha, Unani, Homoeopathy and Naturopathy. In India, about 6,000–7,000 plant species are utilized in traditional, folk and herbal medicines. There are over 200 species of herbs and shrubs in *Piper* genus having various medicinal properties. For example, *Piperlongum* (Long Pepper or Pipli) distributed throughout tropical and semitropical regions of India and other Asian countries is traditionally utilized in the Indian system of medicine for treatment of various ailments.

Herbal medicines/formulations involve the use of crude or processed plants containing several active constituents. Due to the rapid commercialization of herbal medicines, knowledge of phytochemical composition of such crude drugs has become a very crucial aspect for the preparation, safety and efficacy of the herbal products. According to the regulations of the European Medicines Evaluation Agency (EMEA) and Food and Drug Administration (FDA), identification and determination of the active constituent are the crucial prerequisites for the development of modern evidence-based phytomedicine. The use of medicinal herbs or herbal drugs is increasing throughout the world, although one of the main encumbrances in its acceptance globally is the lack of quality control or standardization. Standardization is an essential step for the establishment of consistent pharmacological activity. A reliable chemical profile or simply a quality control program for the production of herbal drugs from medicinal plants can serve the purpose. Standardization of medicinal herbs or herbal products is necessary to manage safety, efficacy and quality, and essential for quality assurance of these drugs.

Recently, mass spectrometry has earned a central role in the field of plant metabolomics. It facilitates efficient analysis of metabolites in the complex matrix of plant extracts due to high sensitivity, selectivity and versatility. LC coupled with tandem mass spectrometry detection (LC-MS/MS) is a sensitive, selective and efficient technique for the comprehensive qualitative and quantitative analysis of plant metabolites. The work contained in this book includes metabolic profiling of *P. nigrum*, *P. longum*, *P. chaba*, *P. mullesa*, *P. umbellatum*, *P. hymenophyllum*, *P. argyrophyllum*, *P. attenuatum*, *P. colubrinum* and *P. galeatum* using DART-MS and quantitative estimation of phytoconstituents using LC-MS. Thirteen bioactive constituents including phenolic acids, flavonoids, propenyl phenol and terpenoid were simultaneously quantified in leaf, root and fruit extracts of different *Piper* species from different geographical regions using UHPLC-QqQ$_{LIT}$-MS/MS.

31/12/2019   **The Authors**

Acknowledgments

The completion of this book is due to the Almighty who blessed us with all the resources required to accomplish this journey. We are glad to have this chance to express our gratitude to people who have been supportive to us at every stage. We are thankful to Dr. K. P. Madhusudanan for his encouragement, guidance and support from beginning to end. His constructive criticism and warm encouragement made it possible for us to bring the work in its present shape. We express our deep sense of gratitude to the Director of CSIR-Central Drug Research Institute (CDRI), Lucknow, India, for his support and Sophisticated Analytical Instrument Facility (SAIF) Division of CSIR-CDRI, Lucknow, India, where all the data were generated.

Authors

Dr. Brijesh Kumar is a Professor (AcSIR) and Chief Scientist of Sophisticated Analytical Instrument Facility (SAIF) Division, CSIR-Central Drug Research Institute (CDRI), Lucknow, India. Currently, he is facility in charge at SAIF Division of CSIR-CDRI. He has completed his PhD from CSIR-CDRI, Lucknow (Dr. R.M.L Avadh University, Faizabad, UP, India). He has to his credit 7 book chapters, 1 book and 145 papers in reputed international journals. His current area of research includes applications of mass spectrometry (DART MS/Q-TOF LC-MS/ 4000 QTrap LC-MS/ Orbitrap MS^n) for qualitative and quantitative analyses of molecules for quality control and authentication/standardization of Indian medicinal plants/parts and their herbal formulations. He is also involved in the identification of marker compounds using statistical software to check adulteration/substitution.

Ms. Surabhi Tiwari completed her master degree in chemistry from the University of Allahabad, India. She has worked in Pharmacognosy Division of NBRI, Lucknow, India, for analysis of herbals using instruments such as HPTLC and HPLC. Currently, she is working as a Senior Research Fellow in SAIF Division under the supervision of Dr. Brijesh Kumar at CSIR-Central Drug Research Institute (CDRI), Lucknow, India. Her current research interest includes phytochemical analysis of medicinal plants.

Dr. Vikas Bajpai completed his PhD from the Academy of Scientific and Innovative Research (AcSIR), New Delhi, India, and carried out his research work under the supervision of Dr. Brijesh Kumar at CSIR-CDRI, Lucknow, India. His research interest includes the development and validation of LC-MS/MS methods for qualitative and quantitative analyses of phytochemicals in Indian medicinal plants.

Dr. Bikarma Singh is working as a Scientist in CSIR-Indian Institute of Integrative Medicine, Jammu, India, and is a recognized Assistant Professor in Academy of Scientific and Innovative Research (AcSIR), New Delhi, India. He graduated as an Honours and Gold Medalist in botany from North-Eastern Hill University, Shillong, India, in 2005, and completed his PhD in botany from Gauhati University, Guwahati, India, Assam and Botanical Survey of India, Shillong, India, in 2012. He authored/co-authored 8 books and published 82 scientific research papers/experimental findings in peer reviewed national and international journals.

List of Abbreviations and Units

°C	degree celsius
µg	microgram
µL	microliter
APCI	atmospheric pressure chemical ionization
API	atmospheric pressure ionization
BPC	base peak chromatogram
CAD	collision activated dissociation
CE	capillary electrophoresis
ce	collision energy
CID	collision induced dissociation
CXP	cell exit potential
Da	dalton
DAD	diode array detection
DP	declustering potential
EP	entrance potential
ESI	electrospray ionization
FDA	food and drug administration
FIA	flow injection analysis
g	gram
GC-MS	gas chromatography-mass spectrometry
GS1	nebulizer gas
GS2	heater gas
h	hour
HPLC	high-performance liquid chromatography
ICH	international conference on harmonization
IS	internal standard
IT	ion trap
kPa	kilopascal
L	liter
LC	liquid chromatography
LOD	limit of detection

LOQ	limit of quantification
LTQ	linear trap quadrupole
m/z	mass-to-charge ratio
mg	milligram
min	minute
mL	milliliter
mM	millimolar
MRM	multiple reaction monitoring
MS	mass spectrometry
ms	millisecond
MS/MS	tandem mass spectrometry
ng	nanogram
NMR	nuclear magnetic resonance
PCA	principal component analysis
PDA	photodiode array
psi	pressure per square inch
QqQ$_{LIT}$	hybrid linear ion trap triple-quadrupole
QTOF	quadrupole time of flight
***r*^{2}**	correlation coefficient
RDA	retro-Diels–Alder
RSD	relative standard deviation
S/N	signal-to-noise ratio
SD	standard deviation
***t*$_R$**	retention time
UHPLC	ultra high-performance liquid chromatography
UV	ultraviolet
WHO	world health organization
XIC/EIC	extracted ion chromatogram

Introduction

1

Since very ancient times, humans have used plants as medicine for their wellness. The use of medicinal plants for healing purposes is as old as mankind. Some of the life-saving drugs in the modern medicine have originated from plants. The plant kingdom is a store house of a variety of organic components, many of them having excellent medicinal properties. Moreover, herbal medicines play an important role in the primary healthcare of a majority of the world's population. The main traditional systems of medicine practiced in India, Ayurveda, Homeopathy, Siddha and Unani, used plant-based medicines. Herbal formulations play an important role in the successful management of various diseases without any side effects. As the use of herbal formulations has increased due to their medicinal properties in recent years, the genus *Piper* has received considerable attention. The widely distributed *Piper* species in tropical and sub-tropical regions of the world, have been used as folk medicine (Parmar et al. 1997; Stöhr et al. 2001). The well-known therapeutic properties of *Piper* genus have made it an important ingredient in Indian system of medicine (ISM). *Piper nigrum* is one of the most widely recognized species of this genus. The presence of several exotic phytochemicals as secondary metabolites is the main reason for the traditional uses of *Piper* species. Different classes of compounds such as alkaloids, amides, benzoic acids, chalcones, chromenes, flavonoids, lignans, long-chain esters, phenolics, phenylpropanoids, pyrones, steroids and terpenoids are reported to be present in *Piper* species (Parmar et al. 1998). Alkaloids are the most important class of compounds in *Piper* species, and a piperamide, piperine (1-piperoylpiperidine), the principal alkaloid occurring in the fruits of *Piper* species, has shown diverse pharmacological activities (Khajuria et al. 1998; Dorman and Deans 2000; Sunila and Kuttan 2004; Lee et al. 2005; Bhat et al. 1987; Desai et al. 2008; Mujumdar et al. 1990; Dhuley et al. 1993; Daware et al. 2000; Bajad et al. 2001b).

1.1 COMMON NAMES OF *PIPER* SPECIES

Piper nigrum is commonly known as Black Pepper in English; Pilpil in Persian; Poivre in French; Schwartze in German; Filfiluswud or Fil-fila-siah in Arabic; Maricham, Maricha, Hapusha, Krishnam, Ooshnam or Valliyam in Sanskrit; Gulmirch or Kalimirch in Hindi; Gol-mirich or Golmorich in Punjabi; Martz in Kashmiri; Vellajung in Bengali; Kalomirich in Gujarati; KaaleMeere in Marathi; Miire in Konkani; Jalook in Assamese; Miriyalu in Telugu; Gurumusi in Manipuri; Golamaricha in Oriya; Kara mirch in Sindhi; Milagu in Tamil; Kuru-mulaka or Kuru-milagu in Malayalam; and Filfil Siyah or Kalimirich in Urdu (Dhanalakshmi et al. 2017).

1.2 DISTRIBUTION OF *PIPER* SPECIES

The genus *Piper* is of great commercial and economic importance in pharmaceutical and food industries. It belongs to the family Piperaceae which consists of mainly herbs (terrestrial and epiphytes), shrubs, vines or trees having six genera with 3,000 species containing alkaloids and aromatic volatile oils. Piperaceae family is distributed in tropical and subtropical regions of the world. In terms of phylogenetic position, Piperaceae is among the diverse assemblage of dicots termed "paleoherbs," and the plants resemble monocots in certain vegetative features such as adaxial prophyll and scattered vascular bundles (Chopra and Vishwakarma 2018). The most common genera are *Piper* and *Peperomia* having 1,000 and 700 species, respectively. *Piper* is one of the most diversified genera within the family Piperaceae, and it occupies a basal lineage position in the angiosperm group. *Piper* includes bushes and herbs that can be found in humid or wet places all around the world, especially in tropical regions of both the hemispheres. *Piper* has the largest number of species found in America (700 species), followed by Southern Asia (300 species), South Pacific (140 species) and Africa (15 species). The genus *Piper* is one of the largest genera of basal angiosperms (TPL 2019) and distributed from sea level to the high ranges of Andes and sub-Himalayan ranges (Royle 1893). It grows wild in the tropical rainforests of India, Indonesia, Malaysia, the Philippines, Sri Lanka and Timor (Chopra and Vishwakarma 2018) (Figure 1.1). In India, 110 species of *Piper* are reported from different regions. The two most important centers of origin of the genus *Piper* in India are the Trans-Gangetic belts and the South Deccan (Hooker 1886; (Purseglove et al. 1981). The foothills of the Western Ghats are

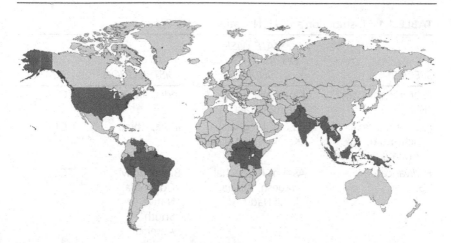

FIGURE 1.1 Worldwide distribution of genus *Piper*.

believed to be the center of origin of *P. nigrum* L., one of the most important medicinal plants in Indian system of medicine. However, *P. nigrum* and a diversity of other *Piper* species are also recorded from Assam, Khasi hills, Mikir hills, lower hills of West Bengal, evergreen forest of the Western Ghats from Konkan to Kerala and Nicobar Islands (Chopra and Vishwakarma 2018).

The distribution patterns of *Piper* species range from locally endemic to widespread. Several species are restricted to a specific center of diversity, whereas some others occur throughout the Neotropics or the Paleotropics (Chopra and Vishwakarma 2018). Most of the *Piper* species are distributed throughout the tropical regions of the earth as shrubs, herbs and lianas, usually growing in the understory of tropical lowland wet forests. *Piper* species as important structural components of the forest understory in the Neotropic belts. In his study on Ayurveda, the Greek physician Hippocrates, known as the father of medicine, referred to *P. longum* for the first time and described it as a medicinal herb rather than spice. The history of black pepper is often confused with that of long pepper, although Theophrastus distinguished the two in the first work of botany more than 2,000 years ago. *P. longum* grows in the western slopes of the Western Ghats of India and is also cultivated in Assam, Andhra Pradesh and Tamil Nadu. The distributional details of *Piper* species from different countries are given in Table 1.1.

In India, the *Piper* distribution is mainly in the Eastern Himalayas and the Southern Deccan. More than 100 species of *Piper* are known, and out of them, 65 species are reported from the northeast *region* only. The Western Ghats are famous for *P. nigrum* L. (black pepper). In 1678, Hendrich Van Rheede described the five types of wild varieties of peppers in his book

TABLE 1.1　Distribution and IUCN Status of *Piper* Species

SPECIES	STATE-WISE DISTRIBUTION IN INDIA	GLOBAL DISTRIBUTION	IUCN STATUS
P. argyrophyllum Miq.	Karnataka, Tamil Nadu, Kerala	India, Sri Lanka	COL
P. attenuatum Buch.-Ham. ex Miq.	Assam, Meghalaya, Kerala, Tamil Nadu	India, China	COL
P. divaricatum G. Mey.	Assam, Arunachal Pradesh, Kerala, Tamil Nadu	Brazil, Bolivia, Columbia; Native to South America	COL
P. galeatum (Miq.) C. DC.	Kerala, Tamil Nadu	–	COL
P. hymenophyllum (Miq.) Wight	Arunachal Pradesh, Kerala, Karnataka, Tamil Nadu	French Guiana	COL
P. longum L.	Assam (cultivated), Meghalaya, Tamil Nadu, Andhra Pradesh	China, Indonesia, Myanmar	COL
P. mullesua Buch.-Ham. ex D.Don	Uttarakhand, Sikkim	China, Nepal, Burma	COL
P. nigrum L.	Assam, Meghalaya Kerala, Tamil Nadu, Karnataka	Brazil, China, Nepal	Endangered
P. retrofractum Vahl	Karnataka	Indonesia, Burma, Malaysia	COL
P. umbellatum L.	Arunachal Pradesh, Karnataka, Kerala	Brazil, China, Philippines	COL

Note: COL-listed in Catalogue of Life.

Hortus Indicus Malabaricus. In 1753, Linnaeus in his book *Species Plantarum* described 17 species from India. Again in 1832, Roxburgh reported seven species of *Piper* from the Indian Peninsula. Seven wild species of *Piper* were reported by Miquel in 1848. In 1869, De Candolle described 52 species of *Piper* from India in his monograph. The first and major work on *Piper* and its species was described by Hooker in 1886 in his famous book *Flora of British*

India, in which he reported forty-five species under six categories, namely, Chavica (Miq.) Hook. f., Cubeba (Raf.) Hook. f., Eupiper C. DC., Heckeria Benth. et. Hook. f., Muldera (Miq.) Benth. et. Hook. f. and Pseudochavica Hook. f and also reported the most difficult genera to *Piper*. Floristic studies have been carried out by various botanists in different parts of India such as the work reported by De Candolle in 1912, 1914 and 1923 from Manipur, Sikkim, Meghalaya and West Bengal; Burkill in 1924–1925 from Arunachal Pradesh; Gamble in 1925 from Western Ghats; from Assam; Deb in 1961 and 1983 from Manipur and Tripura; from Karnataka; Grierson and Long in 1984 from Sikkim and Bhutan; and Ravindran et al. in 1987 from Kerala. Floristic studies on *Piper* are mainly from the Western Ghats of South India. In his Flora of Presidency of Madras, Gamble (1925) reported taxonomic keys to 13 species of *Piper* described the taxonomic and biosystematic studies on *Piper* from the Western Ghats of South India. Ravindran reported taxonomic keys for *Piper* species from the Western Ghats in 1996, in which he has divided the genus into two sections: "Pippali" and "Maricha" on the basis of their inflorescence characteristics like spikes, erect and pendant. The names were taken from Sanskrit corresponding to long pepper and black pepper, respectively. *Piper* species of South India are important from the economic point of view as the major share of its cultivation is from South India. Kerala is famous for the cultivation of *Piper* species, especially black pepper, due to the weather and climatic conditions of the moist deciduous semi-evergreen and evergreen forests of the Western Ghats. Kerala contributes almost 97% of India's total pepper production. Sixteen species of the genus are from South India, and eight species are endemic to the Western Ghats. Hooker in 1886 in his book named *Flora of British India* reported 33 species in which 20 species were from Northeast India, Bhutan, Myanmar, Bangladesh and Nepal. De Candolle in 1912, 1914 and 1923 reported 40 species from Northeast India. Adding to Hooker's record, the total number of species reported from this region is 60.

1.3 BOTANICAL DESCRIPTION OF *PIPER* SPECIES

Piper species is characterized by their alternative leaf pattern, with opposed type of inflorescences axillary, single or compound, floral bract deltoid, triangular, occasionally smooth, glossy, densely clustered flowers on the vertical axis, anthers opening in a vertical, horizontal or oblique plane; pistil 3–5 carpels, smooth fruit, stylish, depressed or truncated. There are many more species that are center specific. Ravindran and Babu (1996, 1997) discussed the morphological characters of different species of *Piper* of South India. The two subgenera from South Indian taxa, Pippali and Maricha, proposed by Ravindran (2000) could be differentiated on the basis of their inflorescence characteristics. Pippali shows spike-like inflorescence, for example, *P. longum* L., whereas in Marichait is hanging or pendant-like inflorescence, for example, *P. nigrum* L. The *Piper* species are grown in shady places and in low and moderate temperatures, with optimum level of moisture in the soil and adequate supply of water. The climber species are grown as an intercrop with trees for support. Dyer et al. in 2004 discussed the vegetative propagation of *Piper* species in four modes, namely, branch repositioning, fragmentation of branches, rhizomes fragmentation and prostration of branches in the soil (Brintnall and Molly 1986).

1.3.1 *Piper argyrophyllum* Miq.

The plant grows as climbing bushy shrub having glabrous leaves. Stems noticeably ridged and furrowed when dried, and leaf petiole size varies, usually 3–3.5 cm long and gets shortest on leaves toward apex of stem. Leaves are chartaceous to thinly coriaceous, ovate-orbicular, 7–21 cm long and 5–7 cm wide, membranous, glandular, usually the lower side of the leaves covered with white hair-like scales; the base is rounded or subcordate and apex cuspidate or mucronate. The inflorescences are drooping in nature and seen as leaf-opposed, and the bracts are with free margins, usually remain adnate to the rachis. Flowers are monoecious; male spikes 8–21 cm, slender; stamens 3 in number; filaments nearly as short as ovoid anthers; anthers ovoid; female spikes 7–9 cm and rachis sparsely hairy around ovoid ovaries; stigmas 3 or 4, slightly curved. Berries are rounded and matured, and turns into black, usually 3–4 mm in diameter.

1.3.2 *Piper attenuatum* Buch.-Ham. ex Miq.

The plant grows as a slender climber. Stems are ridged and furrowed, and petiole size of leaves varies from 3 to 5 cm. Leaves are very thin and dried leaflets, membranous to chartaceous, usually ovate; their size varies from 9 to 15 cm long and 4 to 7.5 cm wide; occasionally small white dots are seen on lower surface of leaflets, whereas upper surface is dark green, glabrous. Inflorescence of flowering twigs is narrow, and filiform varies in size from 10 to 25 cm long; bracts present are obovate to elliptic and usually fall down after fruit maturity. Flowers are with three stamens and one carpel; ovaries are solitary, oblong and stigmas 4-lobed. Berries become rounded when matured; however, immature green berries look oblong 3–4 mm in diameter and taste bitter.

1.3.3 *Piper divaricatum* G.Mey. (Synonym: *P. adenophyllum* Miq., *P. apiculatum* C.DC., *P. colubrina* (Link) Miq., *P. colubrinum* Link)

The plant is a woody lianas, usually rooting from the nodes of stems and petioles 1–1.5 cm long. Leaves vary in shape and size, usually ovate-oblong, 8.5–14.5 cm long and 4.5–8 cm wide, thick and leathery in nature. The base of leaves is rounded or slightly oblique, whereas the apex is acute. Inflorescence is leaf-opposed and sometime becomes as long as leaves. Flowers are monoecious, and bracts present are 3–3.5 cm long and very thin size in broad length; stamens are only 2, present 1 on each side of ovary and anthers reniforms; ovary is globose and stigmas 3 or 4. Berries are sessile, globose, ca. 3 cm diameters, slightly greenish and slowly turn red when ripe.

1.3.4 *Piper galeatum* (Miq.) C.DC. (Synonym: *Muldera galeata* Miq.)

The plant is stout climbers with soft stems and petioles 1–2 cm long. Leaves usually elliptic-ovate, 8–12 cm long and 3–6 cm wide, absence of hairs on the outer surface of the bracts; leaflets are 3–5 ribbed, acute at base and acuminate at apex. Inflorescence looks pinkish, and spikelets are 10 cm long and look drooping. The bracts present are pubescent, having minute hairs and usually formed a cup-like structure. Flowers distantly arranged; male spike slender and shorter than female spike; stamens 2. The fruits are berries, globose, matured ones yellowish red with 3 stigmatic lobes 3.

1.3.5 *Piper hymenophyllum* (Miq.) Wight

The plant is a long twinning climbers having rooting at nodes of stems, and the internodes vary in size from 5 to 8 cm and petioles 0.8–1.5 cm long. Leaves are alternate, elliptic-ovate to lanceolate, vary in size, 6–12 cm long and 2.5–4 cm broad, hairy; the base of the leaves is round to broadly attenuate, and the apex is acuminate. Male inflorescence slender, 5–12 cm long, 0.1–0.2 cm diameter, where female ones 6–8 cm long, 0.2–0.25 cm diameter; stamens usually 2 in number with rounded anthers; ovary minute and sessile, style recurved. Berries are globose, small in size, 0.2–0.4 cm diameter, usually becomes tetra-angular when dried having persistence styles.

1.3.6 *Piper longum* L.

The plants are climbers, grow 3–10 m tall or more depending on climate; young parts are finely pubescent; the stems are often flexuous, and the petioles of the leaflets are 2–9 cm long. Leaves are ovate to oblong, 6.5–10 cm long and 3.5–9.5 cm wide, papery, densely glandular, base cordate, slightly incurved; veins 7, apical pair partly closely parallel to midvein, reaching leaf apex. Inflorescences are leaf-opposed, recurved. Flowers are dioecious; male spikes are longer than female spikes, and vary in size, 4–5 cm long and 2–3 mm wide; bracts are suborbicular, slightly cuneate and 1–2 mm wide; and stamens 2 in number; filaments are short and anthers ellipsoid; female spikes 1.5–2.5 cm long and 3–4 cm wide; bracts 0.9–1 mm in diameter; ovaries are ovoid and stigmas 3 in number, ovoid. Berries are globose, 1–2 mm diameter, yellowish orange, and matured dried berries are brown.

1.3.7 *Piper mullesua* Buch.-Ham. ex D.Don (Synonym: *P. guigual* Buch.-Ham. ex D.Don)

The plants are slender-branched lianas, and stems are 1.5 cm thick, whereas branches are entirely glabrous. Leaves alternate, coriaceous, elliptic, 7–14.5 cm long, 3.5–5.5 cm wide, oblique to rounded at base and acuminate at apex. There are two pairs of prominent lateral ribs; the anterior-most ribs emerge one-third above the leaf base, and the second ones emerge from the base. Inflorescence erect, slightly cylindrical in male, 4–4.5 cm long; the female spikes small, oblong, white in color; the fruit-bearing spikes 1–1.5 cm long and 5–7 mm wide; the bracts are orbicular and peltate; the stamens are 2 in number; the filaments short and the anther lobes single reniform, usually attached transversely at the tip of the filament; the carpel is one and ovary ellipsoid, style represented by a constriction; the stigmas 3-lobed. Berries spherical-obovate, pungent and when eaten gives a burning sensation.

1.3.8 *Piper nigrum* L.

The plants are woody lianas; usually, stems are rooting from the nodes; the petioles are grooved, 1–1.5 cm long. Leaves as shown in Figure 1.2 are fleshy coriaceous, usually ovate to elliptic, vary in size from 9 to 11 cm long and 4 to 6 cm wide; the base of the leaflets is rounded to oblique, whereas the apex is acuminate; the veins are 7 in number, usually two pairs are basal and one

P. nigrum   P. chaba   P. longum

FIGURE 1.2 *Piper* species plants.

pair arising 1–2 cm slightly apart from the base reaching to the apex of the leaves. Inflorescence with male and female flowers together varies in size from 5 to 13 cm long and 0.3 to 0.4 cm diameter; peduncles of the inflorescence are 1–1.5 cm long; stamens are 2 in number male spike and stigmas are 3 in number in female. Berries are globose, sessile, arranged loosely on rachis.

1.3.9 *Piper retrofractum* Vahl (Synonym: *P. chaba* Hunter, *P. officinarum* C.DC.)

The plants are woody climbers growing in open areas. Stems are stout and rooting from the nodes helps in climbing high trees, and petiole of the leaflets is 0.8–1.2 cm long. Leaves coriaceous, oblong to ovate or lanceolate, 10–17 cm long and 4–6 cm wide, glabrous; the base of leaflets iscordate or oblique, whereas the apex is acuminate; the veins are either one or two pairs from basal, and the remaining arise alternately from the midrib. Inflorescence straight, not dropping like other *Piper* inflorescence, usually 3–5.5 cm long and 0.5–0.7 cm diameter; the bracts are orbicular; stamens are 2 in number in male spike, and stigmas are 3 in number in female spike stout. Berries are rounded, usually embedded on rachis, red color when matured.

1.3.10 *Piper umbellatum* L.

The plants are scandent under-shrubs and vary in height from 50 to 80 cm tall, and their stems are 1.5–2.0 cm diameter; the leaflets are petiolated; the petioles are 2–4 cm long. Leaves are thin, chartaceous, and symmetrical, and vary in size from 20 to 30 cm long, 4 to 5 cm wide and the undersurface is glaucous, whereas the upper surface is green when fresh and turns blackish-gray when gets dried. The base of leaves is cordate, and the apex is cuspidate; usually two pairs of veins arise from the base and two pairs from the midribs raised at the both surfaces. Inflorescence axillary and umbellate, opposite to the leaves, usually 10–16 cm long. Berries are globose, 2–3 mm diameter, and turn black when dried.

1.4 BRIEF REVIEW ON CHEMISTRY OF *PIPER* SPECIES

There are nearly 600 compounds isolated from *Piper* species including 145 alkaloids/amides, 47 lignans, 70 neolignans, 89 terpenes and other miscellaneous compounds such as steroids, kawa pyrones, piperolides, chalcones, dihydrochalcones, flavonoes and flavanones (Parmar et al. 1997). *Piper* species are rich in essential oils which are found in roots, stems, leaves, fruits and seeds. These essential oils contain all classes of volatile organic compounds, but the compositions vary depending on several factors (Salehi et al. 2019). These phytochemicals are reported from leaves, stems, roots, fruits and inflorescences of the plants. Principal chemical constituents reported from *Piper* species are phenolics, flavonoids, alkaloids, amides and steroids, lignans, neolignans, isoflavones, isoflavanones, flavones and flavanones, C-glucosyl, pterocarpanes, propenylpiperloids, chalcones, dihydrochalcones and terpenes (Ahmad et al. 2012; Acharya et al. 2012; Taqvi et al. 2008; Manoharan et al. 2009). But the active principle and the principal alkaloid of *Piper* species is piperine (Salehi et al. 2019). The other chemical constituents reported from different *Piper* species are 3β, 4α-dihydroxy-1-(3-phenylpropanoyl)piperidine-2-one, (2E,4E,14Z)-6-hydroxyl-*N*-isobutyleicosa2,4,14-trienamide, Coumaperine, (4-hydroxy-3-methoxyphenyl)-2Epentenoyl piperidine, Piperolactam A,1-[1-oxo-5 (3,4-methylenedioxyphenyl)-2E, 4Epentadienyl] –pirrolidine, (R)-(-) –turmerone, Octahydro-4-hydroy-3alpha-methyl-7methylene-alpha-(1-methylethyl)-1H-indene-1methanol, Aphanamol I, Bisdemethoxycurcumin, Demethoxycurcumin, Longumosides A, Longumosides B, Erythro-1-[1-oxo-9(3,4-methylenedioxyphenyl)8,9-dihydroxy-2E-nonenyl]-piperidine, Threo-1-[1-oxo-9(3,4-methylenedioxyphenyl)8,9-dihydroxy-2E-nonenyl]-piperidine 3β, 4α-dihydroxy-2-piperidinone,5,6-dihydro-2(1H)-pyridinone Piperlongumide(1) [*N*-isobutyl-19-(3′,4′methylenedioxyphenyl)-2E, 4E nonadecadienamide], 1-(3,4-methylenedioxyphenyl)-1E tetradecene, Piperlongimin A [2E-*N*-isobutylhexadecenamide], 2E, 4E-*N*-isobutyl-octadecenamide, Piperlongimin B [2E-octadecenoylpiperidine], 2E, 4E-*N*-isobutyl-dodecenamide, 2E, 4E, 12E, 13-(3,4-methylenedioxyphenyl)trideca-trienoic acid isobutyl amide,

Piperine, Pellitorine, N-[(2E,4E)-Decadienoyl]-piperidine, N-Isobutyl-2E,4E-undecadienamide, Piperlonguminine, Piperanine, N-[(2E,4E)-Tetradecadienoyl] piperidine, N-Isobutyl-2E, 4E-hexadecadienamide, Pipercallosine, (2E,4E,12Z)-N-Isobutyl-octadeca-2,4,12trienamide, N-Isobutyl-2E, E-octadecadienamide, Dehydropipernonaline, Pipernonatine, (E)-9-(Benzo[d][1,3]dioxol-5-yl)-1-(piperidin-1yl)non-2-en-1-one, 1-(2E,4E,12E)-Octadecatrinoylpiperidine, Retrofractamide B, (2E, 4E, 14Z)-N-Isobutyleicosa-2,4,14trienamide, N-isobutyl-2E, 4E-decyldecadienamide, (2E,4E,10E)-N-11-(3,4Methylenediox yphenylhmdecatrienoylpiperidine,1-[(2E,4E,14Z)-1-Oxo-2,4,14-eicosatrienyl] piperidine, Guineensine, (2E,4E,14Z)-N-Isobutyldocosa-2,4,14trienamide, (2E,4E,12E)-13-(Benzo[d][1,3]dioxol-6-yl)-1(piperidin-1-yl)trideca-2,4,12-trien-1-one, (2E,4E,13E)-14-(Benzo[d][1,3]dioxol-6-yl)-Nisobutyltetradeca-2,4,13-trienamide, Brachyamide B, Dihydropiperlonguminine, Piperdardine, Retrofractamide A, Piperchabamide D, N-isobutyl-2E, 4E-dodecadienamide, Piperchabamide B, 13-(1,3-Benzodioxol-5-yl)-N-(2-methylpropyl) (2E,4E)-tridecadienamide, Piperchabamide C, 1-[(2E,4E)-1-oxo-2,4-hexadecadienyl]piperidine, 2,2-Dimethoxybutane, 2 Hydroxy myristic acid, β-Myrcene, N-methyl 1 octadecanamine, Piperazineadipate, 2-Nonynoic acid, Dodecanal, bis(2-ethylhexyl) ester, 2-Amino-4-hydroxypteridine-6-carboxylic acid, Piperlongumine, Palmitic acid, 1,8-cineole, Luwsone, Cis-Decahydronaphthalene, Piperonylic acid, Hypnon, Moslene, Cymol, Methyl hydrocinnamate, Hexahydropyridine (PIP), Piperonal, Isobutylisovalerate, Tridecane (TRD), Pentadecane (MYS), N-Heptadecane, N-Nonadecane (UPL), Tridecylene, Heptadecene, Pentadecene, Nonadecene, tetradecadiene-1,13, Linalool, Cyclopentadecane, Beta-Bisabolene, Sesamol, Sesamin, p-Amino-o-cresol, 2,4-Dimethoxytoluene, D-Camphor (CAM), Cis-2-Decalone, Piperitenone, Nopinene, Alpha-Pinene, Isodiprene, (R)-linalool, N-(2,5-dimethoxyphenyl)-4-methoxybenzamide, Anethole, Isoeugenol, (3S)-3,7-dimethylocta-1,6-dien-3-yl] propanoate, Terpinen-4-ol, Alpha-Farnesene, Farnesene, alpha-humulene, Isocaryophyllene, p-Ocimene, 8-Heptadecene, 9,17-OCTADECADIENAL (Z), Cyclodecene, 1,4,7,-Cycloundecatriene, 1,5,9,9-tetramethyl-, Z,Z,Z-Isoborneol, (Z)-caryophyllene, cis-.beta.-Elemenediastereomer, N-Isobutyl-2,4-icosadienamide, (E,E,E)-11-(1,3-Benzodioxol-5-yl)-N-(2methylpropyl)-2,4,10-undecatrienenamide, Epieudesmin, Valencene,(1S, 5S)-1-isopropyl-4methylenebicyclo[3.1.0] hexane, (5S)-5-[(1R)-1,5-dimethylhex-4-enyl]-2methylcyclohexa-1,3-diene, (E)-5-(4-hydroxy-3-methoxy-phenyl)-1piperidino-pent-2-en-1-one, (3R,8S,9S,10R,13R,14R,17R)-17-[(2R,5S)-5ethyl-6-methylheptan-2-yl]-10,13-dimethyl2,3,4,7,8,9,11,12,14,15,16,17-dodecahydro-1Hcyclopenta[a]phenanthren-3-ol (ZINC03982454), Delta-elemene, (2R, 4aR, 8aR)-2-methyldecalin, (1R, 5R, 7S)-4,7-dimethyl-7-(4-methylpent-3enyl)bicyclo[3.1.1]hept-3-ene, Calarene, Bisdemethoxycurcumin, 1,4-cadinadiene, Tricyclene, Alpha-Cubebene,

Piperundecalidine,   3-Phenylundecane,   4-[(1-Carboxy-2-methylbutyl)amino]-2(1H)pyrimidinone, Bicyclo[3. 2. 2]non-6-en-3-one, Cedryl acetate, Isolongifolene epoxide, N-Isobutyleicosa-2(E), 4(E), 8(Z)-trienamide, Pisatin, Tetradecahydro-1-methylphenanthrene,   Undulatone,   Copaene,   Linalool, Sylvatine, beta-Cubebene, Caryophyllene oxide, alpha-Cedrene, Fargesin, Piperolactam A, Sesamin, 2-Phenylethanol, Zingiberene, Piperine, Chavicine, Piperidine, Piperetine, Resin, Brachyamide B, Dihydro-pipericide, (2E,4E)-N-Eicosadienoyl-pereridine,   N-trans-Feruloyltryamine,   N-Formylpiperidine, Guineensine, Pentadienoyl as Piperidine, (2E,4E)Nisobuty-ldecadienamid, Isobutyl-eicosadienamide, Tricholein, Trichostachine, Isobutyl-eicosatrienamide, Isobutyl-octadienamide, Piperamide, Piperamine, Piperettine, Pipericide, Piperine, Piperolein B (Mgbeahuruike et al. 2017; Sengupta 1987).

1.5 BRIEF REVIEW ON PHARMACOLOGY AND BIOLOGICAL ACTIVITIES OF *PIPER* SPECIES

Piper species have been reported to exhibit various pharmacological activities such as insecticidal, antibacterial, antiinflammatory, antiplatelet, antihypertensive, hepatoprotective, antithyroid and antitumor activities (Iqbal et al. 2016; Khushbu et al. 2011; Kirtiker and Basu 2003; Yang et al. 2002; Ghoshal et al. 2002; Singh et al. 2009; Scott et al. 2008; Kaou et al. 2008; Karsha and Bhagya Lakshmi 2010; Vaghasiya et al. 2007; Tsai et al. 2005; Munshi and Ljungkvist 1972) (Table 1.2).

1.5.1 Insecticidal and Acaricidal Activity

Pipernonaline and piperoctadecalidine, the two alkaloids present in the essential oil of *P. longum*, exhibited insecticidal and insect-repellant activities against five species of arthropod pests (Kokate et al. 1980; Park et al. 2002).

1.5.2 Antifungal Activity

Essential oil of *P. longum* possesses broad-spectrum antifungal activity. In an *in vivo* study fruit, essential oil exhibited excellent antifungal activity against *Botrytis cineria*, *Erysiphe graminis*, *Pyricularia oryzae*,

TABLE 1.2 Biological Activities of *Piper* Species

S. NO.	SPECIES	PART USED	ACTIVITY	REFERENCE
1.	P. longum	Essential oil	Insecticidal and insect-repellant	Kokate et al. 1980; Jeong et al. 2002
2.	P. longum	Ethanolic extract	antiulcer	Agrawal et al. 2000
3.	P. nigrum	Ethanolic extract	Gastric emptying inhibitory	Bajad et al. 2001b
4.	P. longum	Ethanolic extract	Bioavailability enhancers	Atal et al. 1981
5.	P. longum	Ethanolic extract	Antisnake venom	Shenoy et al. 2013
6.	P. longum	Alcoholic extract of fruit	Antiplatelet	Das et al. 1998
7.	P. longum	Fruit extract	Coronary vasodilation	Shoji et al. 1986
8.	P. longum	Essential oil	Antifungal	Lee et al. 2001
9.	P. longum	Methanolic extract of fruit	Antiamoebic	Ghoshal et al. 2002; Ghoshal et al. 1996; Sawangjaroen et al. 2004
10.	Piper species	Ethanolic extract of fruit	Adulticidal	Choochote et al. 2006
11.	Piper species	Ethanolic extract of fruit	Antiobesity	Lee et al. 2005
12.	Piper species	Ethanolic extract of fruit	Larvicidal	Chaithong et al. 2006; Yang et al. 2002
13.	Piper species	Ethanolic extract of fruit	Antidepressant	Seon et al. 2005; Song et al. 2007; Lee et al. 2008
14.	Piper species	Alcoholic extract of fruit	Anticancer and Antitumor	Anuradha et al. 2004; Pradeep et al. 2002; Bezerra et al. 2006

(Continued)

TABLE 1.2 (Continued) Biological Activities of *Piper* Species

S. NO.	SPECIES	PART USED	ACTIVITY	REFERENCE
15.	*Piper* species	Alcoholic extract of fruit	Antiasthmatic	Kulshresta et al. 1969; Kulshresta et al. 1971
16.	*Piper* species	Aqueous extract of fruit	Antidiabetic	Nabi et al. 2013; Manoharan et al. 2007
17.	*Piper* species	Alcoholic extract of fruit	Hypocholesterolemic	Wang et al. 1993; Wu and Bao 1992
18.	*Piper* species	Alcoholic extract of fruit	Hepatoprotective	Koul and Kapil 1993; Christina et al. 2006
19.	*Piper* species	Aqueous extract of root	Analgesic	Vedhanayaki et al. 2003
20.	*Piper* species	Alcoholic extract of fruit	Antiapoptotic	Yadav et al. 2014; Natarajan et al. 2006
21.	*Piper* species	Ethanolic extract of fruit	Antiinflammatory	Kumar et al. 2005; Choudhary 2006; Stohr et al. 2001
22.	*Piper* species	Ethanolic extract of fruit	Immunomodulatory	Mananvalan and Singh 1979
23.	*Piper* species	Aqueous extract of fruit	Antiarthritic	Yende et al. 2010
24.	*Piper* species	Methanolic extract of fruit	Protective myocardial	Mishra 2010
25.	*Piper* species	Root	Antifertility	Lakshmi et al. 2006; Garg 1981; Munshi et al. 1972; Munshi et al. 1972
26.	*Piper* species	Ethanolic extract of fruit	Radioprotective	Sunila and Kuttan 2004

Phytophthora infestans, *Puccinia recondita* and *Rhizoctonia solani* (Lee et al. 2001; Ghoshal et al. 2002; Ghoshal et al. 1996). Pipernonaline (alkaloid) isolated from the hexane fraction of *P. longum* was found potent against *P. recondita* (Sawangjaroen et al. 2004). *P. longum* extracts exhibited good antifungal activity against *Aspergillus niger* (Ab Rahman et al. 2014).

1.5.3 Antiamoebic Activity

In an *in vivo* study, *P. longum* (fruit), *P. sarmentosum* (root) and *Quercus infectoria* (nut gall) methanolic extracts were tested for their antiamoebic effects against *Entamoeba histolytica*. The combination of plant extract and metronidazole was found effective in the treatment of caecal wall ulcerations caused by *E. histolytica* (Ghoshal et al. 2002). *Acacia catechu* (resin), *Amaranthus spinosus* (whole plant), *Brucea javanica* (seed), *P. longum* (fruit) and *Q. infectoria* (nut gall) exhibited *in vitro* inhibition of *Blastocystis hominis* (Ghoshal et al. 1996). In another study, *P. longum* fruit and root extracts exhibited antiamoebic activity (Sawangjaroen et al. 2004). Piperine obtained from *P. longum* and ethanolic extract of *P. longum* can cure caecal amoebiasis (Choochote et al. 2006; Ghoshal et al. 2002). Aqueous extract of *P. longum* fruit exhibited giardicidal activity (Atal et al. 1981).

1.5.4 Antimicrobial Activity

P. longum extracts exhibited good antibacterial activity against *Bacillus megaterium*, *Escherichia coli*, *Salmonella albus*, *S. typhi* and *Pseudomonas aeruginosa* (Lokhande et al. 2007), whereas aqueous extract doesn't exhibit any antibacterial activity. *P. longum* also possesses antitubercular activity (Zaveri et al. 2010; Chaithong et al. 2006; Naika et al. 2010).

1.5.5 Effect on Respiratory System

Piperine, a compound isolated from *P. longum*, exhibited antagonistic respiratory depression in mammals and amphibia induced by morphine or pentobarbitone, and increases the hypnotic response in mice. Piperine and nalorphine isolated from *Piper* sp. reversed morphine-induced respiratory depression in rats, whereas unlike piperine, nalorphine antagonizes morphine-induced analgesia in rats. Medullary stimulant factors were present in the petroleum ether extract of *P. longum* (Parganiha et al. 2011; Kulshresta et al. 1969;

Kulshresta et al. 1971; Zaveri et al. 2010). Piplartine, an alkaloid isolated from *P. longum* extract, suppresses the ciliary movements of esophagus in frog (Nabi et al. 2013).

1.5.6 Antiasthmatic Activity

P. longum milk extract reduces the passive cutaneous anaphylaxis in rats and also protects guinea pigs from bronchospasm (Kokate et al. 1980).

1.5.7 Antidiabetic Activity

P. longum ethanolic extract exhibits antihyperglycemic, antilipid peroxidative and antioxidant activities (Gaurav and Tripti 2013; Nabi et al. 2013; Manoharan et al. 2007), while indigenous herbal formulation increases appetite (Yadav et al. 2014).

1.5.8 Hypocholesterolemic Activity

Methyl piperine isolated from *P. longum* significantly decreases the total serum cholesterol and total cholesterol to high-density lipoprotein cholesterol ratio in rats, whereas, oil of *P. longum* inhibits the elevation of hepatic cholesterol in hypercholesterolemic mice (Singh et al. 2018).

1.5.9 Antioxidant Activity

Piper species, such as *P. nigrum* L., *P. betle* L., *P. guineense* Schumach, *P. umbellatum* L., *P. pellucidum* L., *P. cubeba* L. and *P. sarmentosum* Roxb show strong antioxidant potentials (Natarajan et al. 2006). *P. betle* L., *P. walichii* (Miq) Hand Mazz. and *P. betloides* C.DC collected from Northeast India also show antioxidant activity (Tamuly et al. 2015). It has been shown that the antioxidant activity of *Piper* species due to the presence of phenolic acids, flavonoids like quercetin, catechin, coumaric acid and protocatecuic acid (Tamuly et al. 2015). Amrita Bindu, a polyherbal formulation made from the extract of *Piper* plants along with some other plants, is used as an antioxidant. The ingredients of this polyherbal formulation are *P. longum*, *P. nigrum*, *Zingiber officinale*, *Cyperus rotundus* and *Plumbago zeylanica*. The antioxidant strength of the ingredients of this polyherbal formulation was investigated and found in the following order: *P. nigrum* > *P. longum* > *C. rotundus* > *P. zeylanca* > *Z. officinale* (Natarajan et al. 2006).

1.5.10 Analgesic Activity

P. longum roots exhibited weak opioid type analgesia but significant non-steroidal antiinflammatory drug (NSAID) type of analgesic activity in rats (Vedhanayaki et al. 2003).

1.5.11 Antiinflammatory Activity

Antiinflammatory activity was exhibited by the fruit decoction of *Piper* sp. in rat paw edema induced by carrageenin (Iwamoto et al. 2015; Kumari et al. 2012).

1.5.12 Immunomodulatory Activity

Piperinic acid, an active constituent of *P. longum*, decreases the number of lymphocytes (CD4+ and CD8+ T cells) and cytokine levels in sensitized Balb/C mice, while in another study, alcoholic extract of its fruits and piperine was found to be cytotoxic (Devan et al. 2007). Pipplirasayaba, an Ayurvedic preparation containing long pepper as one of its active constituents, activates the macrophages (Agrawal et al. 2000).

1.5.13 Anticancer Activity

Piper sp. ethanolic extract protects the cell surface and membrane integrity in hamster buccal pouch carcinogenesis induced by 7, 12-dimethylbenz[a] anthracene (DMBA) (El Hamss et al. 2003). Piperine inhibits metastasis of the B16F-10 melanoma cells and reduces the tumor nodule formation along with the reduction in collagen hydroxyproline, hexosamine and uronic acid content, and exhibits chemopreventive effects (Pradeep et al. 2002; Selvendiran et al. 2004). Apart from piperine, piplartine, another alkaloidal amide, isolated from *Piper* exhibits *in vitro* antitumor activity. Piperine exhibits antiapoptotic, antioxidative and restorative properties against mutagenic response, and it can be useful in immunocompromised conditions (Rather et al. 2018).

1.5.14 Antidepressant Activity

P. longum ethanolic fruit extract, piperine and piperidine possess antidepressant activity (Lee et al. 2005; Li et al. 2007), while in another *in vitro* study, piperine decreases the brain derived neurotrophic factor (BDNF) mRNA expression in hippocampal neurons (Lee et al. 2008).

1.5.15 Antiulcer Activity

Mahakasyaya, a decoction made up of ginger, *Ferula asafoetida* and *P. longum*, was reported to inhibit the gastric ulcers in rats (Agrawal et al. 2000). Piperine inhibited the gastric emptying of solids and liquids in mice which is independent of gastric acid and pepsin secretion (Bajad et al. 2001a).

1.5.16 Effect on Reproductive System

Pippaliyadi vati, an Ayurvedic contraceptive, a combination of *P. longum* fruits benzene extract and *Embeliaribes* berries methanolic extract, can inhibit the pregnancy of animals (Lakshmi et al. 2006; Bajad et al. 2001a). The treated female rats gave births to low-weight, smaller-size progenies (Bajad et al. 2001a). Pippaliyadi vati does not have any adverse effect on reproductive performance and postnatal development (Munshi et al. 1972). Piperine decreases the mating performance and fertility in Swiss albino mice (Munshi et al. 1972) and showed decrease in intratesticular testosterone concentration and increased serum gonadotropins level.

1.5.17 Bioavailability Enhancement

Piperine induces the cell membrane dynamics and permeability and synthesis of cytoskeletal proteins. The change in the cell membrane permeability enhances the permeability and bioavailability of other drugs such as diclofenas sodium, vasicine and curcumin (Atal et al. 1981; Pattanaik et al. 2006; Singh et al. 2005; Khajuria et al. 2002; Kim et al. 2009; Kim et al. 2010).

1.5.18 Hepatoprotective Activity

P. longum ethanolic fruit extract and piperine both inhibit liver fibrosis. Ethanolic fruit extract improves regeneration process, while piperine reduces enzymatic leakage of glutathione pyruvate transaminase (GPT) and alkaline phosphatase (AP), lipid peroxidation and depletion of glutathione-stimulating hormone (GSH) and total thiols (Manavalan and Singh 1979).

1.6 TRADITIONAL MEDICINAL USES OF *PIPER* SPECIES

Traditional medicinal knowledge mainly comes from the traditional healers whose knowledge is transferred from one generation to the next generation. Traditional healers are named according to the method used for the treatment; for example, herbal healers use plants or natural products to cure the diseases, blowing healers use blowing procedure to treat the small injuries, and masseuses use massage oil or massage process to treat muscle problems. The ethnobotanical studies reveal the importance of *Piper* in various fields such as food industry, pharmaceutical industry and herbal industry (Manoj et al. 2004). The genus *Piper* is globally used as a spice and condiment, and also known worldwide for their medicinal properties. In Ayurvedic system of medicine, it has been described as Katurasam (pungent taste), Katuvipakam (conversion into pungent taste), ushmaveeryam (hot), guna (properties), vata (nature of wind energy), kapha (nature of water energy), pitta-haram (heat energy), ruksham (rough), lagu (light) and Madhura vipaka (Specific digestion), and used in various disease conditions such as Krimi (parasitic disease), shwasa (dyspnoea), Kshaya (pulmonary tuberculosis), Pliharoga (spleen disorders), Vishamajvara (intermittent fever), Arsha (piles), Urustambha (stiffness of thigh), Vatavyadhi (nervous diseases), Nidranasha (insomnia) and Grahani (dysentery). More than 30 formulations have been made, in which Panchakolchurna, Dashmulashatpalghrita, Vasa Avaleha (Chaudhary et al. 2015a), Vyaghri haritaki Avaleha (Chaudhary et al. 2015b) Kanakabindvarishta (Shingadiya et al. 2015), Mahasudarshan churna (Rajopadhye et al. 2012) and Balachaturbhadra churna (Chaudhary et al. 2015b) are the common ones. Acharya charaka has included it in Agryaushadhi ascribed to adding it as Dipaniya Pachaniya-Anaha prashamananam. In Indian traditional system of medicine, the fruits of *P. longum* and *P. nigrum* are used in respiratory tract diseases, such as cough, bronchitis and asthma. These fruits are also given in insomnia and epilepsy. They are also helpful to clear obstruction of bile duct and bladder, and used as digestive, appetizer, alterative, carminative, laxative and tonic. It is hematinic and useful in treating anemia and chronic fevers, and improving memory (Chopra and Vishawakarma 2018). It is applied locally on muscular pains and inflammation. The fruit possesses tonic, digestive, stomachic, antiseptic, emmenagogue and abortifacient properties. *P. longum* is given with honey in doses of 5–10g for indigestion, dyspepsia, flatulent colic, chronic bronchitis, chest affections and asthma. It is also very useful in treating enlarged spleen, gout and lumbago. The fruits of *P. nigrum* are used as condiment, and they act as rubefacient and disinfectant when applied externally (Martins et al. 1998). Fruit is vermifuge and used after childbirth to check post-partum hemorrhage. Besides fruits, the roots of several *Piper* species such as *P. longum* and *P. umbellatum* are used as stimulant, and traditionally used for curing snakebites and scorpion sting. For contraception, Ayurveda uses *P. longum* combined with *Embeliaribes* and borax as pippaliyadi yoga (Balasinor et al. 2007). A formulation containing a mixture of long pepper root (*P. longum*), black pepper (*P. nigrum*) and ginger (*Z. officinale*) in equal parts in several formulations is prescribed as remedy for treatment of catarrh and hoarseness. The stem of *P. chaba* is used to alleviate post-delivery pains in mothers, the root is used to treat asthma and bronchitis, the fruits, which are carminative and stimulant, are used to treat asthma, bronchitis, inflammation and haemorrhoids (Taufiq-Ur-Rahman et al. 2005). Traditionally, *Piper* species are boiled in combination with ginger, mustard oil, buttermilk and curd to be used as a liniment for sciatica and paralysis. According to Balsubramanian et al. (2007), the *Piper* species act as an appetizer, aphrodisiac, expectorant and rejuvenator, and purgative, antiasthmatic, stomachic and antipyretic agent. Besides, most of the *Piper* species are helpful in curing skin diseases, polyuria, abdominal lump, piles, splenomegaly, colic and rheumatoid arthritis (Anonymous 2005). A fermented decoction called Pippali arista used for the treatment of asthma, cough, anorexia and piles is made from long pepper (*P. longum*), lodhra (*S. racemosa*), black pepper (*P. nigrum*), grapes (*Vitis vinifera*) and *Cissampelos pareira*. A combination of fruit powder of black pepper and long pepper is used in the treatment of coma and drowsiness. Traditionally, the fruit of long pepper and the flowers of *Calotropis gigantea* are used for curing asthma. A decoction made from fruits of *P. longum* and *Adhatoda vasica* is used for getting relief from cough. A paste made from long pepper and neem leaves (*Azadiracta indica*) is mixed with cow's milk and ginger essential oil and applied on the scalp to prevent hair loss. Few ethnobotanical applications of traditionally used *Piper* species in India are described in Table 1.3.

TABLE 1.3   Ethno-botanical Applications of *Piper* Species by Indigenous Tribal Communities in India

SPECIES	PART USED	MODE OF USAGE	DISEASE	TRIBE (STATE)	REFERENCE
P. argyrophyllum	Seed, leaves, fruits	Fresh leaves and fruits are crushed and directly applied on forehead	Headache, weakness	Muthuvan tribe (Kerala)	Jose (2013)
P. attenuatum	Leaves	Fresh leaves are crushed to extract juice and taken orally to cure disease.	Urinary problems	Adi tribe (Arunachal Pradesh)	Jeyaprakash et al. (2017).
P. longum	Fruits, root	Fruits are powdered mixed with jaggery and ginger powder, and then the powder is boiled in water and taken thrice daily before food to cure disease.	Malaria, Bodyache	Jaintia tribe (Northeast India)	Sajem and Gosai (2006).
P. mullesua	Fruits	Powdered fruits are mixed with honey, and taken orally to cure disease.	Rheumatis, cough and bronchit's problems	Apatami, Mongpa, Sinpho, Padam and Ildu tribes (Arunachal Pradesh)	Khongsai et al. (2011)
P. nigrum	Seeds	Decoction from the seeds and leaves are used	Cough, bronchitis, torsillitis	Tagin, Hill Miri (now Nyshi) and Galo tribes (Arunachal Pradesh)	Murtem and Chaudhry (2016)

1.7 COMMERCIALLY AVAILABLE PRODUCTS OBTAINED FROM VARIOUS *PIPER* SPECIES

Plant-, animal- and mineral-based medicines used singularly or in combination are the basis of Ayurvedic or traditional medical practices. More than 30 dosage forms are usually prescribed by Ayurvedic physicians. Churna, vati, gutika, awaleha, asvas-arishtas, ghritas and tilas are some of the dosage forms. In earlier days, the physicians themselves used to handle the preparation of formulations, but now, they depend on the marketed products manufactured by Ayurvedic pharmacies. Trikatu, a Sanskrit word meaning three acrids, is an Ayurvedic preparation of *P. nigrum*, *P. longam* and *Z. officinalis*. According to the *Handbook of Domestic Medicine and Common Ayurvedic Remedies*, 370 listed formulations contain either trikatu or its individual ingredients (Majumdar et al. 1978). Various formulations containing *Piper* species available as commercial products are listed in Table 1.4.

1.8 PHYTOCHEMICAL ANALYSIS

The fruits, seeds, leaves, branches, roots and stems of *Piper* plants are rich in essential oils (Mgbeahuruike et al. 2017). Essential oils from *Piper* species have been analyzed by GC and GC-MS (Martins et al. 1998; Liu et al. 2007). Eighty compounds were profiled from ten *Piper* species cultivated in Hainan Island, China, after headspace–solid-phase microextraction (HS–SPME) with gas chromatography–mass spectrometry (GC-MS) (Hao et al. 2018). Besides essential oils, *Piper* species also contain other polar constituents, and hence, HPTLC and HPLC methods have been used for the identification and quantification of constituents from the genus *Piper* (Variyar and Bandyopadhyay 1994; Hamrapurkar et al. 2011; Wood et al. 1988; Hazra et al. 2018; Rajopadhye et al. 2011; Rao et al. 2011; Kamal et al. 2016). LC-MS has also been used extensively for the qualitative and quantitative determinations of the phytoconstituents of *Piper* species (Friedman et al. 2008; Scott et al. 2005a,b; Kikuzaki et al. 1993; Gu et al. 2018; Sruthi and Zachariah et al. 2010, 2016; Sun et al. 2007). Direct analysis in real time–mass spectrometry (DART-MS), an ambient ionization mass spectrometry technique, can analyze materials in their native form (solids, liquids or gases) eliminating the need for sample

TABLE 1.4 Various Products Available in the Market Obtained from *Piper* Species

S. NO.	SPECIES	DISEASES	NAME	COMPANY
1.	P. longum	Treats cough and cold conditions	Dabur Sitopaladi Churna	Dabur
2.	P. longum	Building immunity	Dabur Chyawanprash	Dabur
3.	P. longum	Helps in relieving pain and hoarseness in throat	Dabur Khadiradi Gutika	Dabur
4.	P. longum	Helps in curing anemia	Dabur Lauhasava	Dabur
5.	P. longum	Helps in treating congestion	Dabur Vasavaleha	Dabur
6.	P. longum	Helps in relieving indigestion problems	Dabur Triphala Churna	Dabur
7.	P. longum	Helps in treating asthma and bronchitis	Dabur Shwaasamrit	Dabur
8.	P. longum	Improves gastric function	Himalaya Trikatu	Himalaya
9.	P. longum	Helps in improving cardiovascular functions	Himalaya Herbals Abana 60	Himalaya
10.	P. nigrum	Reduces the stress level	Himalaya Tentex Forte	Himalaya
11.	P. nigrum	Improves the digestion	Himalaya Gasex	Himalaya
12.	P. longum	Protects the liver	Himalaya Biopure	Himalaya
13.	P. nigrum	Helps in relieving digestion problems and enhancing the metabolism	Pippali Churna	Patanjali
14.	P. nigrum	Helps in building immunity	Baidyanath chyawanprash	Baidyanath
15.	P. longum	Helps in relieving digestion problems	Mahasankh Vati	Baidyanath

(Continued)

TABLE 1.4 (Continued)   Various Products Available in the Market Obtained from *Piper* Species

S. NO.	SPECIES	DISEASES	NAME	COMPANY
16.	P. longum	Helps in building the immunity	Maha Sudarshan Churna	Zandu
17.	P. nigrum	For chronic nightfall in men and adolescents	Baidyanath swapandosh	Baidyanath
18.	P. chaba	Helps in treating congestion and cough	Kantakari Avaleha	Dabur
19.	P. chaba	Helps in curing menstrual disorder	Lodhrasava	Dabur
20.	P. chaba	Helps in enlarging prostate gland	Baidyanath Prostaid	Baidyanath
21.	P. chaba	Helps in relieving stomach acidity	Divya Kumaryasava	Patanjali
22.	P. longum	Improves stomach problems	Divya Ashwagandharishta	Patanjali
23.	P. nigrum	Improves immunity, nourishes body tissues	Mrita SanjivaniSura	Baidyanath
24.	P. nigrum	Helps in treating asthma and relieving cough and cold	Vasarist	Baidyanath
25.	P. nigrum	Helps in gynecological conditions	Panchasav	Baidyanath
26.	P. nigrum	Helps in relieving all types of worm	Vidangasav	Patanjali
27.	P. nigrum	Helps in treating viral infection	Mahasudarshan Ghan Vati	Patanjali
28.	P. longum	Helps in treating cardiovascular problems	Hridyamrit Vati extra power	Patanjali
29.	P. nigrum	Helps in curing leucorrhea and pelvic inflammatory disease	Lukol	Himalaya
30.	P. chaba	Useful in treating all types of swelling, edema and inflammation	Shothari Mandura	Baidyanath
31.	P. chaba	Useful in treating fever during pregnancy	Garbhapal Ras	Baidyanath *(Continued)*

TABLE 1.4 (Continued) Various Products Available in the Market Obtained from *Piper* Species

S. NO.	SPECIES	DISEASES	NAME	COMPANY
32.	*P. longum*	For proper digestion	Gaisantak Bati	Baidyanath
33.	*P. nigrum*	Helps in treating heart diseases, intestinal worms, hernia	Kankayan Vati	Baidyanath
34.	*P. longum*	For proper digestion, abdominal pain, loss of appetite	Agnimukh Churna	Baidyanath
35.	*P. chaba*	For all types of swelling	Shothari Mandur	Baidyanath
36.	*P. chaba*	Helps in treating uterus disorder	Pradarantak Lauh	Baidyanath
37.	*P. nigrum*	Helps in treating cancer	Torcumin tablets	Himalaya
38.	*P. longum*	Helps in relieving mental stress, anxiety	Geriforte	Himalaya
39.	*P. longum*	Specially for infants to cure gastrointestinal disorder	Bonnisan	Himalaya
40.	*P. longum, P. nigrum*	For making toothpaste	Hi Ora-K	Himalaya
41.	*P. longum, P. nigrum*	Helps in treating asthma, cough and cold	Talisadi churna	Dabur
42.	*P. nigrum, P. longum, Z. officinalis*	Helps in relieving indigestion, dyspepsia and cough, and curing other jugular diseases	Trikatu churna	Patanjali, Baidyanath

preparation (Cody et al. 2005). Psychoactive *Piper* species, *P. methysticum* and *P. betle*, have been examined by DART-MS fingerprinting followed by chemometric analysis leading to the differentiation and identification of these two species (Lesiak et al. 2016).

Standardization is a critical step in herbal medicine as it checks variations of phytoconstituents and ensures consistency in the quality of herbal products (Casazza et al. 2011; Kim et al. 2011; Heinrich et al. 2012). Continuing our work on Indian medicinal plants and their bioactivities, we have carried out phytochemical analysis of a number of *Piper* species. Metabolic profiling of *Piper* species (*P. nigrum*, *P. longum* and *P. chaba*) was carried out using DART-MS (Bajpai et al. 2010, 2012; Chandra et al. 2014). Quantitative determination and comparative analysis of the phytoconstituents were carried out in different plant parts (fruits and leaves) of ten *Piper* species using UPLC-ESI-MS/MS (Chandra et al. 2015).

Metabolic Profiling of *Piper* Species by Direct Analysis Using Real Time Mass Spectrometry Combined with Principal Component Analysis

2

2.1 PLANT MATERIAL AND CHEMICALS

Plant materials (fruit, leaf and root of *Piper nigrum*, *P. chaba* and *P. longum*) were collected from the plants grown in the herbal garden of Jawaharlal Nehru Tropical Botanic Garden and Research Institute (TBGRI) Palode, Thiruvananthapuram, India. Voucher specimens, *P. nigrum* TBGT-57944, *P. chaba* TBGT-51838 and *P. longum* TBGT-50930, were deposited in TBGRI, Palode, Thiruvananthapuram, India. For direct analysis in real time mass spectrometry (DART-MS) measurements, crude plant parts were used directly as samples. The plant samples were thoroughly washed with tap water and distilled water in order to remove any foreign particle attached to its surface and kept in oven to dry at 40°C.

2.2 OPTIMIZATION OF DART-MS ANALYSIS

The mass spectrometer used was a JMS-100 TLC (AccuTof) atmospheric pressure ionization time-of-flight mass spectrometer (JEOL, Tokyo, Japan) fitted with a DART ion source. The mass spectrometer was operated in positive-ion mode with a resolving power of 6000 (full-width at half-maximum). The orifice 1 potential was set to 28 V, resulting in minimal fragmentation. The ring lens and orifice 2 potentials were set to 13 and 5 V, respectively. Orifice 1 was set to a temperature of 100°C. The RF ion guide potential was 300 V. The DART ion source was operated with helium gas flowing at approximately 4.0 L/min. The gas heater was set to 300°C. The potential on the discharge needle electrode of the DART source was set to 3,000 V; electrode 1 was set to 100 V and the grid 250 V. Freshly cut pieces of plant samples were positioned in the gap between the DART source and mass spectrometer for measurements. Data acquisition was from m/z 10 to 1050. Fifteen repeats for each sample were run to ascertain the reproducibility of the analysis. Exact mass calibration was accomplished by including a mass spectrum of neat polyethylene (PEG) glycol (1:1 mixture of PEG 200 and PEG 600) in the data file. m-Nitrobenzyl alcohol was also used for calibration. The mass calibration was accurate to within 0.002u. Using the Mass Center software, accurate mass measurement and the elemental compositions could be determined on selected peaks. PCA was carried out using Statistica windows version 7.0 (Stat Soft, Inc., Tulsa, OK, USA) statistical analysis software.

2.3 DART-MS ANALYSIS OF *PIPER* SPECIES

The discrimination analysis among the medicinal herbs of the family Piperaceae had previously been carried out for the identification and characterization of phytoconstituents by performing tedious methods of extraction and sample preparation using various chromatographic methods (Rameshkumar et al. 2011). Ambient desorption mass spectrometry has not been applied for the inter-species classification of raw *Piper* species in herbal drugs. DART mass spectra were measured using different parts of the selected *Piper* species, namely, *P. longum*, *P. nigrum* and *P. chaba*. Representative DART mass spectra of the fruit, leaf and root of *P. longum*, *P. nigrum* and *P. chaba* are shown in Figure 2.1. The spectra showed characteristic patterns (fingerprints) for each species.

Based on the published literature, some of the expected phytochemicals (Lee et al. 2005; Bhat and Chandrasekhara 1987) identified in these plants by DART-MS are shown in Table 2.1. These components are directly ionized from the plant part during measurement and appeared as protonated molecular ions in the resulting spectra as shown in Figure 2.1. The major piperamides observed in the DART-MS of *Piper* species correspond to the protonated molecular ions of piperine ($C_{17}H_{19}NO_3$) at *m/z* 286.1438 (Okwute and Edharevba 2013), pellitorine ($C_{14}H_{25}NO$) at *m/z* 224.2009 (Wu et al. 2004), guincensine ($C_{24}H_{33}NO_3$) at *m/z* 384.2533 (Tsukamoto et al. 2002), dipiperamide ($C_{34}H_{38}N_2O_6$) at *m/z* 571.2803 (Tsukamoto et al. 2002), pipernonaline ($C_{17}H_{27}NO_3$) at *m/z* 342.2064 (Tsukamoto et al. 2002), piperlonguminine ($C_{16}H_{19}NO_3$) at *m/z* 274.1438 (Wu et al. 2004) and piperettine ($C_{19}H_{21}NO_3$) at *m/z* 312.1594 (Okwute and Edharevba 2013). One of the characteristic compounds from the fruits of *Piper* species is the amide, Piperine (Bhat and Chandrasekhara 1987; Parmar et al. 1998; Khajuria et al. 1998; Dorman and Deans 2000; Sunila and Kuttan 2004; Lee et al. 2005; Bezerra et al. 2008b) present in *P. nigrum* and *P. chaba* in high abundance, whereas it was of low abundance in *P. longum*. It was also highly abundant in the root and leaf of *P. nigrum*, whereas much lesser amounts were detected in the root and leaf of *P. chaba* and *P. longum*, thus showing clear distinction. Another peak found in the DART mass spectra of fruit and root of *P. longum* *m/z* 224.2009, [M+H]+ ($C_{14}H_{25}NO$) is a characteristic amide compound corresponding to pellitorine.

Another previously reported amide, dipiperamide A ($C_{34}H_{38}N_2O_6$) at *m/z* 571.2803 [M+H]+ (Kozukue et al. 2007), was identified in the DART-MS of fruit and root of *P. nigrum* and *P. chaba* but absent in *P. longum*. Pipernonaline

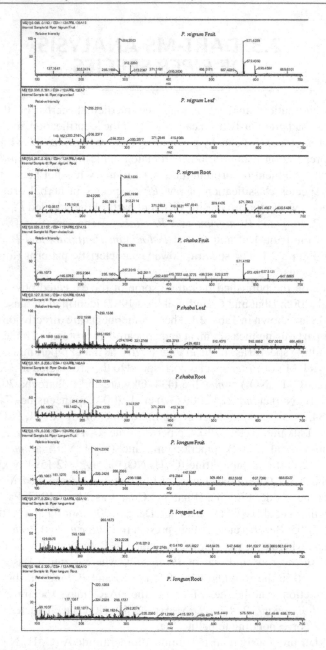

FIGURE 2.1   DART-TOF-MS spectrum of fruit, leaf and root of *P. nigrum*, *P. chaba* and *P. longum*. (Reproduced from Ref. Chandra et al. 2014 with permission from Royal Society of Chemistry.)

TABLE 2.1 DART-MS Accurate Mass Measurement of Phytochemicals in Fruit, Leaf and Root of *P. nigrum*, *P. chaba* and *P. longum*

S. NO.	COMPOUNDS	M. FORMULA	[M+H]+	ERROR (MMU)	P. NIGRUM			P. CHABA			P. LONGUM		
					F	L	R	F	L	R	F	L	R
1	Piperonal	$C_8H_6O_3$	151.039	−0.52	+	−	−	−	−	−	−	−	−
2	(-)-Zingiberene	$C_{15}H_{24}$	205.1951	−0.60	−	+	−	+	−	−	+	−	−
3	N-Cinnamoyl Piperidine	$C_{14}H_7NO$	216.1383	−1.41	−	−	+	−	+	−	−	−	−
4	Sarmentine	$C_{14}H_{23}NO$	222.1857	0.53	+	+	+	−	−	+	−	−	+
5	Pellitorine	$C_{14}H_{25}NO$	224.2009	−2.23	+	−	−	+	−	−	−	−	+
6	Piperchabamide A	$C_{14}H_{15}NO_2$	230.1184	0.32	−	+	−	−	+	−	+	−	−
7	Trichostachine	$C_{16}H_{17}NO_3$	272.128	−0.67	+	−	+	+	+	−	+	−	−
8	Piperlonguminine	$C_{16}H_{19}NO_3$	274.1438	−0.52	+	−	+	+	+	+	+	−	−
9	Piperine	$C_{17}H_{19}NO_3$	286.1438	−0.52	+	+	+	+	+	+	+	+	+
10	Piperchabamide E	$C_{17}H_{21}NO_3$	288.1599	0.18	−	+	+	−	+	+	−	−	+
11	5-[3,4-(Methylenedioxy)phenyl]-pent-2-ene piperidine	$C_{17}H_{19}NO_4$	302.1387	−0.53	+	−	−	+	−	+	−	−	−
12	Piperettine	$C_{19}H_{21}NO_3$	312.1594	1.05	+	+	−	−	−	−	−	−	−
13	Piperdardine	$C_{19}H_{23}NO_3$	314.1751	2.01	−	+	+	−	−	−	−	−	−
14	Piperlongumine	$C_{17}H_{19}NO_5$	318.1336	−0.74	+	−	−	−	−	−	−	−	+
15	Retrofractamide A	$C_{20}H_{25}NO_3$	328.1907	1.27	+	−	−	−	+	−	−	−	−
16	Retrofractamide C	$C_{22}H_{27}NO_3$	330.2064	0.52	+	−	−	−	−	−	−	−	−

(Continued)

TABLE 2.1 (Continued)   DART-MS Accurate Mass Measurement of Phytochemicals in Fruit, Leaf and Root of *P. nigrum*, *P. chaba* and *P. longum*

S. NO.	COMPOUNDS	M. FORMULA	[M+H]+	ERROR (MMU)	P. NIGRUM			P. CHABA			P. LONGUM		
					F	L	R	F	L	R	F	L	R
17	Dehydropipernonaline	$C_{21}H_{25}NO_3$	340.1907	1.27	+	–	–	+	–	–	–	–	–
18	Pipernonaline	$C_{17}H_{27}NO_3$	342.2064	0.52	–	–	+	+	–	–	–	–	–
19	Piperleine B	$C_{21}H_{29}NO_3$	344.222	0.41	+	–	–	–	–	–	+	–	–
20	Pipericide	$C_{22}H_{29}O_3$	356.222	0.57	+	–	–	–	–	–	–	–	–
21	(2E, 4E, 14Z)-*N*-Isobutyleicosa-2,4,14-trienamide	$C_{24}H_{43}NO$	362.3417	2.29	+	–	–	+	–	–	–	–	–
22	Piperchabamide B	$C_{23}H_{31}NO_3$	370.2377	0.52	–	–	–	+	–	+	–	–	–
23	Guineensine	$C_{24}H_{33}NO_3$	384.2533	0.16	+	–	–	–	–	–	–	+	–
24	Piperchabamide C	$C_{25}H_{33}NO_3$	396.2533	0.57	–	–	–	–	–	+	–	–	–
25	Dipiperamide A	$C_{34}H_{38}N_2O_6$	571.2803	–1.13	+	–	+	+	–	–	–	–	–

F-fruit, L-leaf, R-root, + = present, – = absent.
Source: Reproduced from Ref. Chandra et al. 2014 with permission from Royal Society of Chemistry.

$(C_{17}H_{27}NO_3)$ at m/z 342.2064, [M+H]$^+$, was present in the root of *P. nigrum* and *P. chaba*, whereas it was not detected in the root of *P. longum*. Thus, it is clear that DART mass spectra of fruit, leaf and root of *P. nigrum*, *P. chaba* and *P. longum* showed characteristic differences. Piperchabamide A $(C_{14}H_{15}NO_2)$ at m/z 230.1184 [M+H]$^+$ (Morikawa et al. 2004), identified as one of the main components from the leaf of *P. chaba*, was of low abundance in other *Piper* species. Some of the minor peaks observed corresponded to piperonal $(C_8H_6O_3)$ at m/z 151.0390, [M+H]$^+$, trichostachine $(C_{16}H_{17}NO_3)$ at m/z 272.1438, [M+H]$^+$, retrofractamide A $(C_{20}H_{25}NO_3)$ at m/z 328.1907, [M+H]$^+$ and retrofractamide C $(C_{20}H_{27}NO_3)$ at m/z 330.2064, [M+H]$^+$. These constituents were found to discriminate the fruits of *P. nigrum* from other *Piper* species. All the plant parts showed differences in their chemical profile. The distribution of the most common piperamides in fruit, leaf and root of *P. nigrum*, *P. chaba* and *P. longum* reflected in the DART mass spectra and expressed as percent total ionization is shown in Table 2.2. The abundance of piperine, pellitorine and dipiperamide A was high, whereas the other amides were in low abundance.

It is clear that DART-MS results help to do primary screening on the basis of abundances of bioactive molecules to select the most suitable plant part for a particular medicinal application. The anticancer compound, pellitorine, previously isolated (Ee et al. 2009) from roots of *P. nigrum* is present in high abundance in the fruit of *P. longum*, and hence, it can be used as an alternative to *P. nigrum*. The DART-MS fingerprints of *P. nigrum*, *P. chaba* and *P. longum* showed the potentiality and versatility of this technique for the quality control of these medicinal herbs.

TABLE 2.2   (%) Ionization of Peaks in Fruits of *P. nigrum*, *P. chaba* and *P. longum*

M+H	COMPOUNDS	(%) IONIZATION		
		P. NIGRUM	*P. CHABA*	*P. LONGUM*
151	Piperonal	2	Nd	Nd
224	Pellitorine	37	Nd	Nd
286	Piperine	5	43	42
342	Pipernonaline	Nd	Nd	3
384	Guineensine	3	Nd	Nd
571	Dipiperamide A	Nd	24	39

Nd—not detected.

2.4 IDENTIFICATION OF MARKER USING PRINCIPAL COMPONENT ANALYSIS

PCA, factorial analysis and cluster analysis are important and proven techniques for complex data analysis (Li et al. 2009; Rawal et al. 2010). DART-MS fingerprinting in combination with data reduction technique like PCA is a powerful tool for distinguishing and identifying phytoconstituents as markers in foods and medicines. PCA was selected for dimensionality reduction in an attempt to distinguish the characteristic profiles from the DART-MS data and to identify marker phytochemicals that aided in grouping of samples. PCA is an unsupervised procedure that determines the directions of the largest variations in the dataset and the data are generally presented as a two-dimensional plot (score plot) where the coordinate axis represents the directions of the two largest variations. The DART-MS data of *P. nigrum*, *P. chaba* and *P. longum* plant parts were subjected to PCA. The 15 peaks extracted from PCA (*m/z* 169, 195, 224, 238, 272, 289, 302, 312, 371, 447, 461, 559, 571, 557 and 597) were able to discriminate among the species and in between fruit, leaf and root part as shown in the score plot (PC1 vs. PC2) in Figures 2.2 and 2.3, respectively. The score plot of the fruits showed clustering of the data according to the species. Similar clustering and differentiation is also clearly seen in root and leaf. Different parts of the same plant also showed similar clustering. DART-MS followed by PCA seems to be an appropriate method for easy differentiation of species as well as plant parts.

The PCs were able to explain 68.81% of the variance. The PC1 vs. PC2 plot shows a distinctive discrimination between the fruits of *P. nigrum*, *P. longum* and *P. chaba*. The samples of *P. nigrum* are falling in the second quadrant having negative PC score with the main contribution from *m/z* 571. The positive PC2 score is mainly due to peaks at *m/z* 312 and *m/z* 597.

The samples of *P. longum* are along the X-axis, where the main contribution to PC1 score is coming from *m/z* 224, *m/z* 447 and *m/z* 289. There is no contribution of *m/z* 571 in *P. longum*. The *P. chaba* is characterized by the negative scores of both PC1 and PC2. The dominance in *P. chaba* is mainly due to *m/z* 571 and to some extent *m/z* 312. The PC1 and PC2 scores in both are in the range from–1 to –3. But the discrimination between the two is mainly because its contribution in *P. chaba* is higher than that in *P. nigrum*. Peaks at *m/z* 169, *m/z* 195, *m/z* 224, *m/z* 238, *m/z* 289, *m/z* 302 and *m/z* 447 observed in DART mass spectrum of *P. longum* were contributing more indiscrimination from other *Piper* species, i.e., *P. nigrum* and *P. chaba*.

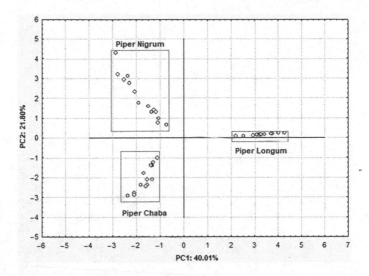

FIGURE 2.2 PC1 vs. PC2 plot shows a distinctive discrimination between the fruits *P. nigrum*, *P. chaba* and *P. longum* on the basis of 15 peaks (169, 195, 224, 238, 272, 289, 302, 312, 371, 447, 461, 559, 571, 557 and 597). (Reproduced from Ref. Chandra et al. 2014 with permission from Royal Society of Chemistry.)

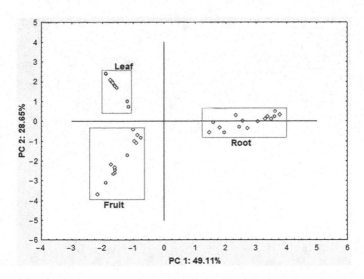

FIGURE 2.3 PCA plot of *P. nigrum* plant parts discrimination on the basis of nine peaks (151.12, 224.22, 236.23, 250.15, 286.2, 447.44, 509.45, 571.42 and 597.41). (Reproduced from Ref. Chandra et al. 2014 with permission from Royal Society of Chemistry.)

Quantitative Determination of Chemical Constituents of Piper Species Using UPLC-ESI-MS/MS

3

3.1 PLANT MATERIAL AND CHEMICALS

Fruits and leaves of *Piper* species were collected from the plants grown in the herbal garden of Jawaharlal Nehru Tropical Botanic Garden and Research Institute (TBGRI), Palode, Thiruvananthapuram, India and Indian Institute of Horticultural Research (IIHR), Hesaraghatta Lake Post, Bangalore, India. Voucher specimens were deposited as per the numbers, namely, **(1)** *Piper argyrophyllum*-50983, **(2)** *P.attenuatum*-50982, **(4)** *P. galeatum*-66414, **(5)** *P. hymenophyllum*-50961, **(6)** *P. chaba*-51838, **(7)** *P. longum*-50930, **(8)** *P. nigrum*-57944, **(9)** *P. mullesa*-50906 and **(10)** *P. umbellatum*-50947 at the TBGRI, Palode, Thiruvananthapuram, India, and **(3)** *P. colubrinum-*IIHRPC1, at IIHR, Bangalore, India.

LC-MS grade acetonitrile and methanol and analytical grade formic acid were purchased from Fluka, Sigma-Aldrich (St. Louis, MO, USA). A Millipore water purification system (Millipore, Milford, MA, USA) provided Milli-Q ultra-pure water. The reference standards (purity ≥ 96%) of apigenin, caffeic acid, curcumin, ferulic acid, kaempferol, palmatine piperine, piperlongumine, protocatechuic acid, rosmarinic acid, ursolic acid and vanillic acid were purchased from Sigma-Aldrich Ltd. (St. Louis),whereas luteolin and quercetin were purchased from Extrasynthese (Genay, France). Piperlonguminine was obtained from Biovision, Inc. (Milpitas, CA, USA). Both curcumin and palmatine were used as IS (internal standards).

3.2 EXTRACTION AND SAMPLE PREPARATION

The washed and dried (room temperature) fruits and leaves of the *Piper* species were ground into powder (40 meshes) separately. The dried powder of each part (5g) was weighed precisely and suspended in methanol (50 mL) and sonicated for 30 min at room temperature using ultrasonic water bath (Bandelin SONOREX, Berlin, Germany), left for 24 hours at room temperature and filtered using Whatman filter paper. The residue was re-extracted thrice with methanol using the same procedure. The combined filtrates were evaporated to dryness under reduced pressure on a rotatory evaporator (Buchi Rotavapor-R2, Flawil, Switzerland) at 40°C. Dried residues (1 mg) were weighed accurately and dissolved in 1 mL of 100% methanol. The solutions were filtered through 0.22 µm syringe filter (Millex-GV, PVDF; Merck Millipore, Darmstadt, Germany). The filtrates were diluted with methanol to final working concentration; 50 µL of both IS were spiked in final working solution and vortexed for 30s, and 5 µL aliquot was injected into the UPLC-MS/MS system for analysis.

A mixed standard stock solution of 0.05 mg/mL containing piperamides (piperine, piperlongumine and piperlonguminine), phenolics (caffeic acid, ferulic acid, protocatechuic acid, rosmarinic acid and vanillic acid), flavonoids (quercetin, kaempferol, apigenin and luteolin) and terpenoid (ursolic acid) was prepared using methanol. The working standard solutions were prepared by diluting the mixed standard solution with methanol to a series of concentrations ranging from 0.5 to 250 ng/mL and were used for plotting the calibration curve after linear regression of the ratios of peak areas of the analytes to those of the internal standard. The standard stock and working solutions were all stored at −20°C until use and vortexed prior to injection (5 μL).

3.3 UPLC-MS/MS CONDITIONS

3.3.1 Instrumental Parameters

An Acquity UPLC system comprising an autosampler and a binary pump (Waters, Milford, MA) equipped with a 10μL loop was interfaced with a hybrid linear ion trap triple-quadrupole mass spectrometer (API 4000 QTRAP™ MS/MS system; AB Sciex, Concord, ON, Canada) equipped with electrospray (Turbo V) ion source. A Waters Acquity BEH C18 (2.1 mm×50 mm, 1.7 μm) column at 30°C was used for separation of the components. Mobile phase consisted of 0.1% formic acid in water (A) and acetonitrile (B), and the components were eluted at the rate of 0.3 mL/min using a gradient program—3% to 30% B from 0 to 3.5 min, 30% to 60% B from 3.5 to 5 min, 60% to 92% B from 5 to 7 min, from 7 to 8 min, followed by a return to the initial condition from 8 to 10.5 min. Sample injection volume was 5 μL. The mass spectrometer conditions in negative ionization mode were ion spray voltage (IS) 4200, turbo spray temperature (TEM) 550°C, nebulizer gas (GS 1), heater gas (GS 2) and curtain gas (CUR) at 20 psi, and in positive ionization mode were ion spray voltage (IS) 5500, turbo spray temperature (TEM) 550°C, nebulizer gas (GS 1) 50 psi, heater gas (GS 2) at 50 psi and curtain gas (CUR) at 20 psi. The interface heater was on, nitrogen was used as the CAD and the scan type used was MRM.

3.3.2 Compound-dependent Parameters

The mass spectrometric conditions were optimized by infusing 50 ng/mL solutions of the analytes dissolved in methanol at a flow rate of 10 μL/min using a Harvard "22" syringe pump (Harvard Apparatus, South Natick, MA, USA). For the MRM quantitation, the highest abundance of precursor-to-product ions for each compound was chosen. The dwell times for both the parent and the internal standard were set at 200 ms. For full-scan ESI-MS analysis, the mass range scanned was from *m/z* 100 to 1000. Analyst 1.5.1 software package (AB Sciex) was used for instrument control and data acquisition.

The contents of thirteen bioactive compounds in fruits and leaves of ten *Piper* species were used for PCA using the software STATISTICA 7.0. When the contents of investigated compounds were below the quantitation limit or not detected in the samples, the values were considered zero.

3.4 OPTIMIZATION OF UPLC CONDITIONS

Unless the analytes are having the same molecular weight and similar structure, complete separation of proximate analytes is certainly not required for MS/MS detection. However, incomplete separation may lead to reduced sensitivity of MS detection due to ion suppression and formation of the same precursor/product ions in MRM detection. In this study, piperine (m/z 286) and piperlonguminine (m/z 274) generated the same product ion at m/z 201. Therefore, mobile-phase composition was optimized by using different solvents and adjusting the gradient elution for separation of all the compounds. The greater elution strength of acetonitrile compared to that of methanol can shorten the analysis time, and thus, acetonitrile was selected for the method development. Based on the polarities of piperamides, phenolics, flavonoids and terpenoids in the extracts of *Piper* species samples, an Acquity UPLC BEH C18 (2.1 mm×50 mm, 1.7 μm; Waters) column was selected for their separation. Compared to acetic acid, formic acid was found more effective for ionization under both positive and negative ESI modes. Thus, different concentrations (0.05%, 0.1% and 0.2%) of formic acid were investigated for optimization of the analysis, and 0.1% formic acid was selected for better ionization. Optimized gradient elution with 0.1% formic acid in water and acetonitrile at a flow rate of 0.3 mL/min with the column temperature of 30°C led to the separation of the 13 compounds in a runtime of 8 min. HPLC methods reported previously for the simultaneous determination of piperine, its isomers and piperlonguminine in black, white, green and red pepper fruits were more expensive in terms of solvent consumption of 0.8–1.0 mL/min and time of analysis of 90–110 min (Kozukue et al. 2007; Friedman et al. 2008). The UPLC-ESI-MS/MS method developed in this study is faster and more efficient as the quantitative comparison of 13 selected bioactive compounds comprising piperamides, phenolics, flavonoids and terpenoids in fruits and leaves of *Piper* species was made in run time of 8 min at a flow rate of 0.3 mL/min.

3.5 OPTIMIZATION OF MS/MS CONDITIONS

The compound-dependent MS parameters, such as precursor ion, product ion, DP and collision energy (ce), were carefully optimized for each target compound individually in both positive and negative ion modes by injecting the individual standard solution into the mass spectrometer. It was observed that the investigated piperamides were adequately ionized in the positive ion ESI mode, whereas phenolics, flavonoids and terpenoid were best ionized in the negative ion ESI mode.

3.6 QUANTITATIVE ANALYSIS

3.6.1 MS/MS Spectra and MRM Transitions

The precursor–product ion pairs described here were selected for MRM transitions. Piperine (m/z 286 [M+H]$^+$) and piperlonguminine (m/z 274 [M+H]$^+$) generated the same major fragment ion at m/z 201 in positive ion mode due to the cleavage of the amide group in their structures. A similar fragmentation in piperlongumine at m/z 318 [M+H]$^+$ led to the product ion at m/z 221.1. Protocatechuic acid m/z 153 [M-H]$^-$ and caffeic acid m/z 179 [M-H]$^-$ produced major fragment ions at m/z 109 and 135, respectively, due to loss of CO_2. Vanillic acid (m/z 167 [M-H]$^-$) produced a major fragment ion at m/z 108 due to losses of methyl radical followed by CO_2 (Gonthier et al. 2003). Ferulic acid at m/z 193 [M-H]$^-$ also showed losses of methyl radical and CO_2 resulting in the ion at m/z 134. Rosmarinic acid at m/z 379 [M-H]$^-$ generated a major fragment ion at m/z 161 corresponding to [M-H-180-H_2O]$^-$ or [caffeic acid-H-H_2O]$^-$ (Hossain et al. 2010).

Retro Diels-Alder (RDA) reaction gave rise to the ion at m/z 151 in quercetin, whereas kaempferol yielded a fragment ion at m/z 239 corresponding to [M-CO-H_2O]$^-$ (Zhao et al. 2007). RDA reaction in luteolin at m/z 285 [M-H]$^-$) and apigenin at m/z 269 [M-H]$^-$ yielded product ions at m/z 133 and 117, respectively. MRM with parent and product ion was not suitable for ursolic acid quantitation as the parent ion isolated in Quadrupole 1 (Q1) passed through Quadrupole 2(Q2) without fragmentation and monitored in Quadrupole 3 (Q3) as such (Xia et al. 2011). The internal standards palmatine at m/z 352 $[M+H]^+$ and curcumin at m/z 367 $[M-H]^-$ yielded product ions at m/z 336 and 217, respectively (Feng et al. 2010; Yang et al. 2007). The MS^2 spectra and fragmentations pattern are shown in Figures 3.1 and 3.2.

MRM parameters, DP and ce, were suitably optimized to achieve the most abundant, specific and stable MRM transition for each compound (Table 3.1). A positive/negative switching time of 100 ms was used for the continuous polarity switching between the positive and negative ionization modesin the LC-MS/MS system. The acquisition method built by the software algorithm aligns both polarities at appropriate MRM transitions. This has resulted in the simultaneous determination of both positively and negatively ionized bioactive compounds in shorter analysis time without additional injections into the LC/MS/MS system.

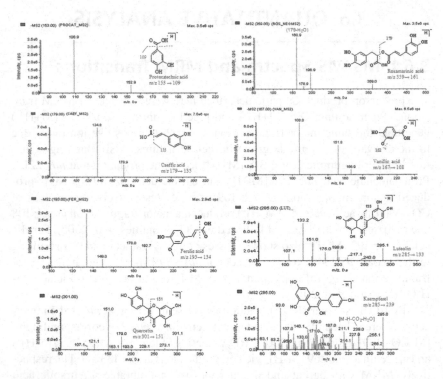

FIGURE 3.1 MS/MS spectra and fragmentation of analytes. (Reproduced from Ref. Chandra et al. 2015 with permission from Elsevier.)

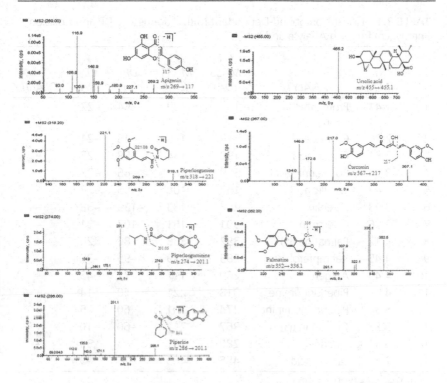

FIGURE 3.2   MS/MS spectra and fragmentation of analytes and IS. (Reproduced from Ref. Chandra et al. 2015 with permission from Elsevier.)

3.6.2 Analytical Method Validation

The proposed UPLC-MRM method for quantitative analysis was validated according to the guidelines of International Conference on Harmonization (ICH, Q2R1) by determining linearity, lower LOD, lower LOQ, precision, solution stability and recovery. MRM extracted ion chromatogram of analytes and IS are shown in Figure 3.3.

3.6.3 Linearity, LOD and LOQ

The IS method, wherein a known amount of a reference compound is added to all samples, was employed to calculate the contents of 13 analytes in *Piper* species. The stock solution was diluted with methanol to different working

TABLE 3.1   List of Compound-Dependent MRM Parameters, DP and Collision Energy (ce) for Each Analyte and IS

S. NO.	RT (MIN)	ANALYTE	Q1 MASS (DA)	Q3 MASS (DA)	DP (V)	CE (EV)	POLARITY
1	1.41	Protocatechuic acid	153	109	−64	−22	−
2	2.06	Caffeic acid	179	135	−48	−21	−
3	2.81	Ferulic acid	193	134	−58	−23	−
4	3.33	Rosmarinic acid	359	161	−65	−22	−
5	3.48	Vanillic acid	167	108	−55	−22	−
6	3.93	Luteolin	285	133	−139	−38	−
7	3.96	Quercetin	301	151	−107	−31	−
8	4.12	Palmatine (IS)	352	336	80	26	+
9	4.45	Kaempferol	285	239	−95	−35	−
10	4.45	Apigenin	269	117	−71	−45	−
11	4.97	Piperlongumine	318	221	40	14	+
12	5.3	Piperlonguminine	274	201	60	25	+
13	5.33	Curcumin (IS)	367	217	−60	−10	−
14	5.41	Piperine	286	201	45	30	+
15	7.27	Ursolic acid	455.1	455	−75	−9	−

+ = [M+H]+, − = [M-H]−
Source: Reproduced from Ref. Chandra et al. 2015 with permission from Elsevier.

concentrations for the construction of calibration curves. The linearity of calibration was performed by the analytes to IS peak area ratios vs. the nominal concentration, and the calibration curves were constructed with a weight factor $(1/x^2)$ by least-squares linear regression. The applied calibration model for all curves was $y = ax + b$, where y = peak area ratio (analyte/IS), x = concentration of the analyte, a = slope of the curve and b = intercept. The LODs and LOQs were measured with S/N of 3 and 10, respectively, as criteria. The results are listed in Table 3.2. All the calibration curves indicated good linearity with correlation coefficients (r^2) from 0.9986 to 0.9999 within the test ranges. The LODs for each analyte varied from 0.02 to 1.34 ng/mL and LOQs from 0.06 to 3.88 ng/mL. As previously reported (Kozukue et al. 2007; Friedman et al. 2008), LOD for piperine was 15–30 ng/mL, but in our study, LODs for all analytes are less than 1.34 ng/mL. Thus, the developed UPLC-ESI-MS/MS method is more sensitive than the previously reported methods.

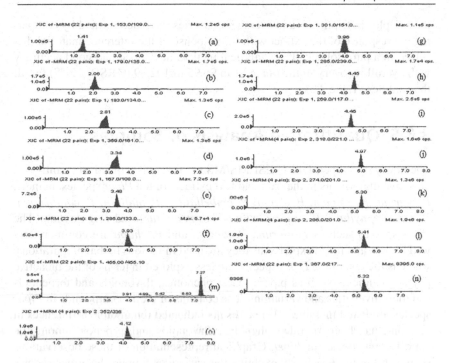

FIGURE 3.3 UPLC-MRM extracted ion chromatogram of analytes: (a) protocatechuic acid, (b) caffeic acid, (c) ferulic acid, (d) rosmarinic acid, (e) vanillic acid, (f) luteolin, (g) quercetin, (h) kaempferol, (i) apigenin, (j) piperlongumine, (k) piperlonguminine, (l) piperine, (m) ursolic acid and IS, i.e., (n) curcumin and (o) palmatine. (Reproduced from Ref. Chandra et al. 2015 with permission from Elsevier.)

3.6.4 Precision, Stability and Recovery

The intra- and inter-day variations chosen to determine the precision of the developed method were investigated by determining thirteen analytes in six replicates in a single day and by duplicating the experiments on three consecutive days. Variations of the peak areas were taken as the measures of precision and expressed as percentage relative standard deviations (%RSD=standard deviation/mean*100). The overall intra- and inter-day precision was not more than 1.66. Stability of sample solutions stored at room temperature was investigated by replicating injections of the sample solution at 0, 2, 4, 8, 12 and 24 h. The %RSD value of stability of the 13 analytes is ≤ 3.14. A recovery test was carried out to evaluate the accuracy of this method. Three different concentration levels (high, middle and low) of the analytical standards were added into the samples in triplicate, and average recoveries were determined by comparing the response of the extract with the response of the reference material dissolved in a pure solvent. The analytical method developed had good accuracy with overall recovery in the range from 96.95 to 102.39 (%RSD ≤ 2.78) for all the analytes (Table 3.2).

3.6.5 Quantitative Analysis of Samples

The newly developed UPLC-ESI-MS/MS method was applied to quantify 13 bioactive ingredients in the fruit and leaf extracts of ten *Piper* species, namely, *P. nigrum*, *P. longum*, *P. chaba*, *P. umbellatum*, *P. mullesa*, *P. colubrinum*, *P. hymenophyllum*, *P. argyrophyllum*, *P. attenuatum* and *P. galeatum*. The *Piper* species such as *P. nigrum*, *P. longum* and *P. chaba* are commercially well known and much explored as they are used in various herbal preparations, whereas the rest of the *Piper* species are less explored in terms of the bioactive phytoconstituents such as piperamides, phenolics, flavonoids and terpenoids and their diversities and variations. The contents of the 13 compounds in *Piper* species are listed in Table 3.3. The results indicated remarkable differences in the contents of piperamides, phenolics, flavonoids and terpenoid among the ten different species of *Piper*. Graphical representation of these observations shown in Figure 3.4 clearly explains the variations among the *Piper* species and between fruits and leaves.

Among the three piperamides, piperine was the major compound in most of the *Piper* species. *P. nigrum* fruits had the highest piperine content (7,220 µg/g), followed by *P. chaba* (5,530 µg/g), *P. longum* (2,150 µg/g) and *P. hymenophyllum* (309 µg/g). Only low piperine content was detected in *P. umbellatum* (2.10 µg/g) and *P. attenuatum* (1.65 µg/g). These results were consistent with the previous reports of determination of piperine (ref). Among the leaves, *P. chaba* (203 µg/g) and *P. mullesa* (271 µg/g) leaves yielded high content of piperine, whereas the lowest content was obtained in *P. umbellatum* (1.09 µg/g). Piperine was below the detection limit in both fruits and leaves of *P. argyrophyllum*. The contents of other piperamides, namely, piperlongumi-nine and piperlongumine, were the highest in fruits of *P. longum* (15,200 µg/g and 162.00 µg/g, respectively). Thus, piperamides could serve as the signature constituents to distinguish among the *Piper* species. As shown in Table 3.3, the total content of phenolics was in close proportion and highest when compared to others in fruits of *P. mullesa* (1,129 µg/g) and leaves of *P. longum* (1,094 µg/g), whereas it was found lowest in both fruits and leaves of *P. chaba* (82 µg/g). The total content of flavonoids was found higher in fruits and leaves of *P. umbellatum* (1,124 µg/g and 625 µg/g, respectively), while components such as kaempferol and luteolin could not be quantified or detected in most

TABLE 3.2 Regression Equation, Correlation Coefficients, Linearity Ranges and Lower LOD and LOQ for 13 Reference Analytes

ANALYTES	REGRESSION EQUATION	R²	LINEAR RANGE NG/ML	LOD NG/ML	LOQ NG/ML	PRECISION RSD (%)		STABILITY RSD, (n = 5)	RECOVERY RSD (%)
						Intra-day (n = 6)	Inter-day (n = 6)		
Protocatechuic acid	y = 8.3201x − 0.0364	0.9993	1–50	0.12	0.36	1.66	1.05	2.78	1.22
Caffeic acid	y = 18.2x + 0.00859	0.9989	1–50	0.16	0.48	0.86	2.63	1.66	1.56
Ferulic acid	y = 2.685x − 0.0209	0.9991	1–50	0.19	0.56	2.22	2.94	1.68	1.55
Rosmarinic acid	y = 12.68x + 0.0914	0.9995	1–50	0.08	0.28	1.72	1.82	2.11	1.27
Vanillic acid	y = 37.795x + 0.1116	0.9998	1–50	0.12	0.40	1.24	1.36	2.15	2.78
Luteolin	y = 3.6406x − 0.0064	0.9995	1–50	0.18	0.52	1.13	1.14	3.14	1.66
Quercetin	y = 9.7543x − 0.0102	0.9991	5–100	0.10	0.33	2.51	1.23	2.86	0.86
Kaempferol	y = 0.7303x − 0.0006	0.9993	10–250	1.34	3.88	0.63	2.36	1.89	0.78
Apigenin	y = 13.426x − 0.0527	0.9992	1–50	0.18	0.52	1.02	1.17	1.66	1.45
Piperlongumine	y = 0.0186x − 0.0076	0.9998	0.5–50	0.06	0.20	0.77	2.25	1.59	2.38
Piperlonguminine	y = 1.133x + 0.221	0.9998	0.5–100	0.04	0.12	3.37	0.93	2.68	1.89
Piperine	y = 0.0063x − 0.0033	0.9999	0.5–100	0.02	0.06	0.88	0.91	2.45	1.87
Ursolic acid	y = 0.6281x − 0.0862	0.9986	5–100	0.62	1.92	1.59	4.09	2.16	1.59

Source: Reproduced from Ref. Chandra et al. 2015 with permission from Elsevier.

TABLE 3.3 Contents (μg/g) of the 13 Active Compounds in Fruits (FR) and Leaves (LF) of ten *Piper* Species (average ± %RSD, n = 3)

| SAMPLE CODE | PIPERAMIDES | | | PHENOLICS | | | | | | | FLAVONOIDS | | | | | TERPENOID |
	PIPERINE	PIPERLONGU-MINE	PIPERLONGU-MINE	TOTAL	PROTOCATE-CHUIC ACID	CAFFEIC ACID	FERULIC ACID	ROSMARINIC ACID	VANILLIC ACID	TOTAL	LUTEOLIN	QUERCETIN	KAEMPFEROL	APIGENIN	TOTAL	URSOLIC ACID
1 FR	bdl	36.0 ± 0.28	bdl	36.0 ± 0.28	50.50 ± 0.02	nd	26.45 ± 0.30	5.30 ± 0.05	3.30 ± 0.01	85.55 ± 0.38	nD	91.1 ± 0.28	nd	5.30 ± 0.69	96.40 ± 0.97	nd
1 LF	bdl	Bdl	bdl	0.00	11.40 ± 0.41	bdl	21.35 ± 1.02	108.0 ± 1.27	nd	140.75 ± 2.70	ND	93.9 ± 1.01	nd	5.90 ± 0.40	99.80 ± 1.41	nd
2 FR	1.65 ± 0.05	40.0 ± 0.66	bdl	41.65 ± 0.71	135.0 ± 1.06	bdl	22.05 ± 0.82	21.20 ± 0.02	3.24 ± 1.01	181.49 ± 2.91	0.78 ± 0.02	22.2 ± 1.22	nd	5.35 ± 1.05	28.33 ± 2.29	nd
2 LF	18.54 ± 1.12	36.30 ± 1.31	bdl	54.84 ± 2.43	88.50 ± 1.21	bdl	72.0 ± 0.44	105.0 ± 0.08	3.69 ± 0.54	269.19 ± 2.27	0.56 ± 0.11	nd	nd	5.90 ± 1.34	6.46 ± 1.45	nd
3FR	bdl	14.10 ± 0.80	nd	14.10 ± 0.80	77.0 ± 0.63	bdl	31.35 ± 0.03	111.0 ± 1.66	3.37 ± 0.81	222.72 ± 3.13	38.2 ± 0.27	885.0 ± 0.48	11 ± 0.03	9.05 ± 0.05	943.25 ± 0.83	2920.0 ± 0.49
3 LF	4.67 ± 0.07	36.80 ± 2.01	nd	41.47 ± 2.08	57.0 ± 2.05	18.0 ± 1.05	59.0 ± 1.02	5.45 ± 0.25	3.25 ± 0.73	142.70 ± 5.10	4.24 ± 0.76	bdl	15.0 ± 0.06	8.25 ± 0.93	27.49 ± 1.75	107.0 ± 0.22
4 FR	10.0 ± 1.03	11.80 ± 1.22	nd	21.8 ± 2.25	5.10 ± 0.09	5.6 ± 0.02	23.0 ± 0.05	4.22 ± 0.04	nd	37.92 ± 0.20	1.33 ± 0.05	10.98 ± 0.64	11.0 ± 0.48	nd	23.31 ± 1.17	1235.0 ± 1.11
4 LF	28.9 ± 0.52	14.50 ± 0.55	nd	43.40 ± 1.07	139.50 ± 0.46	39.35 ± 0.29	42.7 ± 0.57	16.56 ± 0.41	1.71 ± 0.90	239.82 ± 2.63	0.81 ± 0.03	nd	5.65 ± 0.27	6.05 ± 0.06	12.51 ± 0.36	158.0 ± 0.32
5 FR	309.0 ± 3.12	bdl	bdl	309.0 ± 3.12	13.50 ± 1.20	bdl	45.1 ± 1.06	733.0 ± 0.62	1.61 ± 1.02	793.21 ± 3.90	1.03 ± 0.08	527.0 ± 0.80	4.0 ± 0.01	5.95 ± 1.03	537.98 ± 1.92	2420.0 ± 0.50
5 LF	29.30 ± 2.04	10.60 ± 1.77	nd	39.90 ± 3.81	Nd	bdl	27.7 ± 0.10	15.14 ± 0.20	1.65 ± 0.06	44.49 ± 0.36	1.4 ± 0.18	11.44 ± 0.02	16.0 ± 0.04	6.40 ± 0.11	35.24 ± 0.35	810.0 ± 0.44
6 FR	5530.0 ± 3.12	407.0 ± 2.03	bdl	5937.0 ± 5.15	18.50 ± 0.07	15.6 ± 0.03	nd	8.60 ± 0.11	nd	42.70 ± 0.21	3.15 ± 1.01	18.38 ± 0.07	bdl	7.45 ± 0.73	28.98 ± 1.81	nd

(Continued)

TABLE 3.3 (Continued)   Contents (μg/g) of the 13 Active Compounds in Fruits (FR) and Leaves (LF) of ten *Piper* Species (average ± %RSD, $n = 3$)

SAMPLE CODE	PIPERAMIDES			TOTAL	PHENOLICS					TOTAL	FLAVONOIDS				TOTAL	TERPENOID
	PIPERINE	PIPERLONGU-MININE	PIPERLONGU-MINE		PROTOCATE-CHUIC ACID	CAFFEIC ACID	FERULIC ACID	ROSMARINIC ACID	VANILLIC ACID		LUTEOLIN	QUERCETIN	KAEMPFEROL	APIGENIN		URSOLIC ACID
6 LF	203.20 ± 0.98	bdl	65.10 ± 0.21	268.30 ± 1.19	Nd	bdl	30.8 ± 2.03	5.25 ± 1.06	3.23 ± 0.36	39.33 ± 3.45	nd	nd	bdl	5.30 ± 0.48	5.3 ± 0.48	nd
7 FR	2150.0 ± 1.23	15200.0 ± 2.81	162.0 ± 0.09	20512.0 ± 4.13	15.70 ± 0.14	bdl	51.0 ± 0.10	21.20 ± 1.50	nd	87.90 ± 1.74	nd	nd	nd	5.75 ± 0.82	5.75 ± 0.82	1775.0 ± 0.38
7 LF	bdl	43.90 ± 0.11	39.0 ± 0.02	82.90 ± 0.13	15.10 ± 0.64	bdl	15.75 ± 0.23	1060.0 ± 2.01	3.25 ± 1.26	1094.1 ± 4.14	nd	88.8 ± 0.33	nd	5.75 ± 0.29	94.55 ± 0.62	nd
8 FR	7220.0 ± 0.62	62.90 ± 1.09	34.50 ± 0.88	7317.40 ± 2.59	20.10 ± 0.68	bdl	32.30 ± 0.06	5.35 ± 1.92	1.56 ± 0.03	59.31 ± 2.69	nd	nd	8.0 ± 0.22	5.40 ± 0.10	13.40 ± 0.32	193.0 ± 0.05
8 LF	27.60 ± 1.45	48.80 ± 2.33	17.10 ± 1.6	93.50 ± 5.38	18.05 ± 0.02	bdl	55.0 ± 0.12	23.0 ± 0.06	1.59 ± 0.46	97.64 ± 0.66	nd	18.86 ± 1.36	12.0 ± 1.01	6.0 ± 0.51	36.86 ± 2.88	580.0 ± 1.04
9 FR	28.80 ± 0.83	40.0 ± 0.49	nd	68.80 ± 1.32	96.5 ± 0.04	nd	132.5 ± 1.07	897.0 ± 0.71	3.28 ± 0.73	1 29.28 ± 2.55	nd	bdl	bdl	5.20 ± 0.03	5.20 ± 0.03	nd
9 LF	271.0 ± 3.02	36.0 ± 0.53	nd	307.0 ± 3.55	37.55 ± 0.61	20.3 ± 0.49	165.0 ± 0.29	762.0 ± 0.38	3.27 ± 1.40	988.12 ± 3.17	nd	bdl	bdl	6.40 ± 0.17	6.40 ± 0.17	389.0 ± 1.36
10 FR	2.10 ± 0.72	17.10 ± 1.04	64.30 ± 0.04	83.50 ± 1.80	42.6 ± 0.03	7.65 ± 1.13	111.5 ± 0.43	21.40 ± 0.02	nd	183.15 ± .61	1.43 ± 0.17	1410.0 ± 1.27	6.0 ± 0.31	7.0 ± 1.82	1424.43 ± 3.57	8600.0 ± 1.20
10 LF	1.09 ± 0.05	bdl	bdl	1.09 ± 0.05	55.0 ± 0.13	28.25 ± 1.50	131.5 ± 0.06	21.20 ± 1.04	3.2 ± 0.06	239.16 ± 2.79	2.35 ± 0.21	605.0 ± 0.90	1.0 ± 0.55	16.70 ± 0.04	625.05 ± 1.70	3965.0 ± 0.72

nd = not detected; bdl = below detection limit; FR = fruit; LF = leaf; *P. argyrophyllum* (1); *P. attenuatum* (2); *P. colubrinum* (3); *P. galeatum* (4); *P. hymenophyllum* (5); *P. chaba* (6); *P. longum* (7); *P. nigrum* (8); *P. mullesa* (9); *P. umbellatum* (10).

Source: Reproduced from Ref. Chandra et al. 2015 with permission from Elsevier.

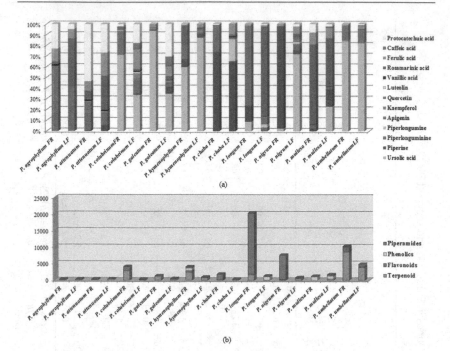

FIGURE 3.4   (a) Graphical representation of 13 compounds plant part of different *Piper* species. (b) Graphical representation of compounds contents in plant part of ten *Piper* species. (Reproduced from Ref. Chandra et al. 2015 with permission from Elsevier.)

of the *Piper* species. Terpenoid (ursolic acid) content was quantitatively found highest in fruits followed by leaves of *P. umbellatum* (8,600 µg/g and 3,965 µg/g, respectively) and lowest in the leaf of *P. colubrinum* (107 µg/g). This characteristic observation helps in differentiating these species among others.

The genus *Piper* is in great demand in herbal industry, and because of the high content of piperamides, *Piper* fruit extracts are incorporated in various formulations. Our findings indicated that piperamides are the signature class of compounds *Piper* species, and their contents are altogether the highest in the fruits of *P. longum* (21,500 µg/g), *P. nigrum* (7,317.40 µg/g) and *P. chaba* (5,937 µg/g) as compared to the other seven species. There were significant differences among the collected plant samples of *Piper* species in terms of individual contents of the investigated bioactive compounds.

3.7 DISCRIMINATION OF *PIPER* SPECIES BY PRINCIPAL COMPONENT ANALYSIS

Modern analytical methods generate large datasets, and in order to interpret such datasets, some dimensionality reduction is essential. An unsupervised pattern recognition method such as PCA can analyze, classify and reduce the dimensionality quite easily (Guo et al. 2014; Nordén et al. 2005). We have carried out PCA of the LC-MS datasets of the ten *Piper* species based on the characteristics of the contents of the thirteen investigated compounds resulting in clear-cut differences among the fruits and leaves of ten *Piper* species. Figure 3.5 shows the score plot obtained by PCA of the contents of the phyto-components in fruits ("FR") and leaves ("LF").

PCA was also run on the combined datasets of fruits and leaves, thus giving a matrix of size 20×13. Zero variation in the quantities (SD = 0) of compounds in the *Piper* plants was indicative of overall consistency in preparation and

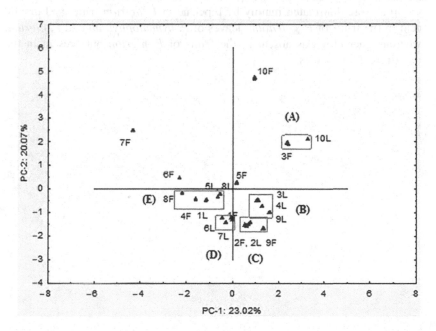

FIGURE 3.5 PCA score plot discrimination of *Piper* species. (Reproduced from Ref. Chandra et al. 2015 with permission from Elsevier.)

experimentation. Wherever the variation occurred randomly due to unknown reasons. The information of the data matrix was contained in five principal components explaining 76.5% of total variation in data. Considering the closeness of species in PCA score plot, all species were grouped into five categories: A, B, C, D and E (Figure 3.5). The clustering of group A was between two species leaves of *P. umbellatum* and fruits of *P. colubrinum*. This was dominated by ursolic acid and quercetin. The species contained in group B have high amount of ursolic acid, and quercetin and piperlongumine were absent in them. The similarity in having a small quantity of vanillic acid in these species is its characteristic. Similar features in the fruits of *P. mullesa* and leaves and fruits of *P. attenuatum* have brought them in the group C. This group was dominated by protocatechuic acid and ferulic acid. The complete absence of piperlongumine and ursolic acid together with an almost equal quantity of vanillic acid and piperlonguminine among all the three species of *Piper* was the distinctive feature of this group. The clustering of group D was mainly due to the availability of apigenin, vanillic acid and piperlonguminine in almost equal amounts. The complete absence of caffeic acid, luteolin, kaempferol and ursolic acid together in this group was also its characteristic. Five species in group E were dominated mainly by piperine in *P. nigrum* fruits and ursolic acid in the fruits of *P. galeatum*, leaves of *P. hymenophyllum* and *P. nigrum*, whereas quercetin was absent in the fruits of *P. nigrum* but present in the remaining four species.

Conclusions

4

This book described the study of *Piper* species using two different techniques, namely, DART-MS and UPLC-ESI-MS/MS. Analysis of three *Piper* species by DART-MS provided a useful method for high-throughput analysis for the identification of bioactive compounds. It can also help in selection of right plant/part, authentication and quality control purposes. The DART-MS technique has been applied first time for the successful profiling of alkaloids and amides in the fruit, leaf and root of *Piper nigrum*, *P. chaba* and *P. longum*. Significant differences in mass spectra were obtained. Statistical analysis of the most abundant ions provided accurate classification of plant samples of *Piper* species. Marker peaks were identified for fruit, leaf and root for discrimination of *Piper* species with the help of PCA.

More complex but sensitive techniques have been employed for determination of secondary metabolites in *Piper* species. Development and validation of an UPLC-MRM method for quantitative analysis of constituents from *Piper* species are described. The developed method was applied for the quantitative analysis of 13 selected bioactive compounds comprising piperamides, phenolics, flavonoids and terpenoid in fruits and leaves of ten *Piper* species in a single run with continuous polarity switching method. PCA was used to compare and evaluate the quality of ten *Piper* species based on the characteristics of the contents of thirteen investigated compounds. The highest content of piperamides in fruits compared to leaf in different *Piper* species clearly suggested the suitability of this part as raw material in the manufacturing of *Piper* based products. Quality and quantity of bioactive constituents are the most important factors for efficacy of drugs and food supplements. The comparative analysis of contents of piperamides, phenolics, flavonoids and terpenoid will felicitate manufacturers to select suitable raw herbs according to their requirements. This method is rapid and sensitive giving a quantitative fingerprint of the constituents and helping in ensuring the quality control of herbal medicinal products.

References

Ab Rahman, Mas Rizal, Fathilah Abdul Razak, and Marina Mohd Bakri. "Evaluation of wound closure activity of *Nigella sativa, Melastoma malabathricum, Pluchea indica,* and *Piper sarmentosum* extracts on scratched monolayer of human gingival fibroblasts." *Evidence-Based Complementary and Alternative Medicine,* Article ID 190342 (2014): 1–9.

Acharya, S. G., A. H. Momin, and A. V. Gajjar. "Review of piperine as a bio-enhancer." *American Journal of PharmTech Research* 2 (2012): 32–44.

Agrawal, A. K., Ch V. Rao, K. Sairam, V. K. Joshi, and R. K. Goel. "Effect of *Piper longum* Linn, *Zingiber officianalis* Linn and *Ferula* species on gastric ulceration and secretion in rats." *Indian Journal of Experimental Biology* 38(2000): 994–998.

Ahmad, Nisar, Hina Fazal, Bilal Haider Abbasi, Shahid Farooq, Mohammad Ali, and Mubarak Ali Khan. "Biological role of *Piper nigrum* L. (Black pepper): a review." *Asian Pacific Journal of Tropical Biomedicine* 2, no. 3 (2012): S1945–S1953.

Anonymous. *The Treatise on Indian Medicinal Plants* (Vol. 1, p. 1013). New Delhi: NISCAIR, 2005.

Anuradha, V., P. V. Srinivas, and J. Madhusudana Rao. "Isolation and synthesis of isodihydropiperlonguminine." *Natural Product Research* 18, no. 3 (2004): 247–251.

Atal, C. K., Usha Zutshi, and P. G. Rao. "Scientific evidence on the role of Ayurvedic herbals on bioavailability of drugs." *Journal of Ethnopharmacology* 4, no. 2 (1981): 229–232.

Bajad, Sunil, K. L. Bedi, A. K. Singla, and R. K. Johri. "Antidiarrhoeal activity of piperine in mice." *Planta Medica* 67, no. 3 (2001a): 284–287.

Bajad, Sunil, K. L. Bedi, A. K. Singla, and R. K. Johri. "Piperine inhibits gastric emptying and gastrointestinal transit in rats and mice." *Planta Medica* 67, no. 2 (2001b): 176–179.

Bajpai, Vikas, Deepty Sharma, Brijesh Kumar, and K. P. Madhusudanan. "Profiling of Piper betle Linn. Cultivars by direct analysis in real time mass spectrometric technique." *Biomedical Chromatography* 24, no. 12 (2010): 1283–1286.

Bajpai, Vikas, Renu Pandey, Mahendra Pal Singh Negi, K. Hima Bindu, Nikhil Kumar, and Brijesh Kumar. "Characteristic differences in metabolite profile in male and female plants of dioecious *Piper betle* L." *Journal of Biosciences* 37, no. 1 (2012): 1061–1066.

Balasinor, Nafisa, Ashima Bhan, Niraja S. Paradkar, Arifa Shaikh, Tarala D. Nandedkar, K. K. Bhutani, and Mandakini Roy-Chaudhury. "Postnatal development and reproductive performance of F1 progeny exposed in utero to an ayurvedic contraceptive: Pippaliyadi yoga." *Journal of Ethnopharmacology* 109, no. 3 (2007): 406–411.

Balasubramanian, G., M. Sarathi, S. R. Kumar, A. S. S. Hameed. "Screening the antiviral activity of Indian medicinal plants against White spot syndrome virus in shrimp." *Aquaculture* 263, (2007): 15–19.

Bezerra, D. P., F. O. Castro, A. P. N. N. Alves, C. Pessoa, M. O. Moraes, E. R. Silveira, M. A. S. Lima, F. J. M. Elmiro, and L. V. Costa-Lotufo. "In vivo growth-inhibition of Sarcoma 180 by piplartine and piperine, two alkaloid amides from Piper." *Brazilian Journal of Medical and Biological Research* 39, no. 6 (2006): 801–807.

Bhat, B. G., and N. Chandrasekhara. "Effect of black pepper and piperine on bile secretion and composition in rats." *Food/Nahrung* 31, no. 9 (1987): 913–916.

Brintnall, Simpson Beryl, and Conner-ogorzaly Molly. *Economic Botany: Plants in Our World*. New York: McGraw-Hill, 1986.

Casazza, A. P., F. Gavazzi, F. Mastromauro, S. Gianì, and D. Breviario. "Certifying the feed to guarantee the quality of traditional food: an easy way to trace plant species in complex mixtures." *Food Chemistry* 124, no. 2 (2011): 685–691.

Chaithong, Udom, Wej Choochote, Kittichai Kamsuk, Atchariya Jitpakdi, Pongsri Tippawangkosol, Dana Chaiyasit, Daruna Champakaew, Benjawan Tuetun, and Benjawan Pitasawat. "Larvicidal effect of pepper plants on Aedes aegypti (L.) (Diptera: Culicidae)." *Journal of Vector Ecology* 31, no. 1 (2006): 138–144.

Chandra, Preeti, Renu Pandey, Mukesh Srivastva, K. B. Rameshkumar, and Brijesh Kumar. "Quantitative determination of chemical constituents of Piper spp. using UPLC-ESI-MS/MS." *Industrial Crops and Products* 76 (2015): 967–976.

Chandra, Preeti, Vikas Bajpai, Mukesh Srivastva, K. B. Ramesh Kumar, and Brijesh Kumar. "Metabolic profiling of Piper species by direct analysis using real time mass spectrometry combined with principal component analysis." *Analytical Methods* 6, no. 12 (2014): 4234–4239.

Chaudhary, S. A., R. K. Shingadiya, K. S. Patel, V. K. Kori, S. Rajagopala, C. R. Harisha, and V. J. Shukla. "Pharmacognostical & pharmaceutical evaluation of balachaturbhadra vati – a well-known drug for pediatric disorders. "*International Journal of Ayurveda and Pharmaceutical Chemistry* 3 (2015a): 192–210.

Chaudhary, Suhas A., S. Patel Kalpana, V. K. Kori, S. Rajagopala, C. R. Harisha, and V. J. Shukla. "Pharmacognostical and pharmaceutical evaluation of Vasa Avaleha—an ayurvedic compound." *IJPBA* 6, no. 1 (2015b): 32–36.

Choochote, Wej, Udom Chaithong, Kittichai Kamsuk, Eumporn Rattanachanpichai, Atchariya Jitpakdi, Pongsri Tippawangkosol, Dana Chaiyasit, Daruna Champakaew, Benjawan Tuetun, and Benjawan Pitasawat. "Adulticidal activity against Stegomyia aegypti (Diptera: Culicidae) of three Piper spp." *Revista do Instituto de Medicina Tropical de São Paulo* 48, no. 1 (2006): 33–37.

Chopra, V. L., and R. A. Vishwakarma. *Plants for Wellness and Vigour* (399 pp). New Delhi: New India Publishing Agency, 2018.

Choudhary, G. P. "Mast cell stabilizing activity of Piper longum Linn." *Indian Journal of Allergy, Asthma and Immunology* 20, no. 2 (2006): 112–116.

Christina, A. J. M., G. R. Saraswathy, S. J. Heison Robert, R. Kothai, N. Chidambaranathan, G. Nalini, and R. L. Therasal. "Inhibition of CCl4-induced liver fibrosis by Piper longum Linn." *Phytomedicine* 13, no. 3 (2006): 196–198.

Cody, Robert B., James A. Laramée, and H. Dupont Durst. "Versatile new ion source for the analysis of materials in open air under ambient conditions." *Analytical Chemistry* 77, no. 8 (2005): 2297–2302.

Daware, Mahadeo B., Arvind M. Mujumdar, and Surendra Ghaskadbi. "Reproductive toxicity of piperine in Swiss albino mice." *Planta Medica* 66, no. 3 (2000): 231–236.

Desai, S. K., V. S. Gawali, A. B. Naik, and L. L. D'souza. "Potentiating effect of piperine on hepatoprotective activity of *Boerhaavia diffusa* to combat oxidative stress." *International Journal of Pharmacognosy* 4 (2008): 393–397.

Devan, Prabhavathy, Sarang Bani, Krishna Avtar Suri, Naresh Kumar Satti, and Ghulam Nabi Qazi. "Immunomodulation exhibited by piperinic acid through suppression of proinflammatory cytokines." *International Immunopharmacology* 7, no. 7 (2007): 889–899.

Dhanalakshmi, D., S. Umamaheswari, D. Balaji, R. Santhanalakshmi, and S. Kavimani. "Phytochemistry and pharmacology of *Piper longum*: a systematic review." *World Journal of Pharmacy and Pharmaceutical Sciences* 6, no. 1 (2017): 381–398.

Dhuley, J. N., P. H. Raman, A. M. Mujumdar, and S. R. Naik. "Inhibition of lipid peroxidation by piperine during experimental inflammation in rats." *Indian Journal of Experimental Biology* 31, no. 5 (1993): 443–445.

Dorman, H. J. D., and Stanley G. Deans. "Antimicrobial agents from plants: antibacterial activity of plant volatile oils." *Journal of Applied Microbiology* 88, no. 2 (2000): 308–316.

Dyer, Lee A., and Aparna D. N. Palmer, eds. *Piper: A Model Genus for Studies of Phytochemistry, Ecology, and Evolution.* New York: Kluwer Academic/Plenum Publishers, 2004.

Ee, G. C. I., C. M. Lim, C. K. Lim, M. Rahmani, K. Shaari, and C. F. J. Bong. "Alkaloids from *Piper sarmentosum* and *Piper nigrum*." *Natural Product Research* 23, no. 15 (2009): 1416–1423.

El Hamss, R., M. Idaomar, A. Alonso-Moraga, and A. Munoz Serrano. "Antimutagenic properties of bell and black peppers." *Food and Chemical Toxicology* 41, no. 1 (2003): 41–47.

Feng, Jin, Wen Xu, Xia Tao, Hua Wei, Fei Cai, Bo Jiang, and Wansheng Chen. "Simultaneous determination of baicalin, baicalein, wogonin, berberine, palmatine and jatrorrhizine in rat plasma by liquid chromatography-tandem mass spectrometry and application in pharmacokinetic studies after oral administration of traditional Chinese medicinal preparations containing scutellaria-coptis herb couple." *Journal of Pharmaceutical and Biomedical Analysis* 53, no. 3 (2010): 591–598.

Friedman, Mendel, Carol E. Levin, Seung-Un Lee, Jin-Shik Lee, Mayumi Ohnisi-Kameyama, and Nobuyuki Kozukue. "Analysis by HPLC and LC/MS of pungent piperamides in commercial black, white, green, and red whole and ground peppercorns." *Journal of Agricultural and Food Chemistry* 56, no. 9 (2008): 3028–3036.

Garg, S. K. "Antifertility effect of *Embelia ribes* and *Piper longum* in female rats." *Fitoterapia* 52, no. 4 (1981): 167–169.

Gaurav, K. S., and V. Tripti. "Antihyperlipidemic activity of seed extract of *Piper attenuatum* in triton X-100 induced hyperlipidemia in rats." *Journal of Chemical and Pharmaceutical Research* 5, no. 12 (2013): 1370–1373.

Ghoshal, Sheela, B. N. Krishna Prasad, and V. Lakshmi. "Antiamoebic activity of *Piper longum* fruits against *Entamoeba histolytica* in vitro and in vivo." *Journal of Ethnopharmacology* 50, no. 3 (1996): 167–170.

Ghoshal, Sheela, and V. Lakshmi. "Potential antiamoebic property of the roots of *Piper longum* Linn." *Phytotherapy Research: An International Journal Devoted to Pharmacological and Toxicological Evaluation of Natural Product Derivatives* 16, no. 7 (2002): 689–691.

Gonthier, Marie-Paule, Laurent Y. Rios, Marie-Anne Verny, Christian Rémésy, and Augustin Scalbert. "Novel liquid chromatography-electrospray ionization mass spectrometry method for the quantification in human urine of microbial aromatic acid metabolites derived from dietary polyphenols." *Journal of Chromatography B* 789, no. 2 (2003): 247–255.

Gu, Fenglin, Guiping Wu, Yiming Fang, and Hongying Zhu. "Nontargeted metabolomics for phenolic and polyhydroxy compounds profile of pepper (*Piper nigrum* L.) products based on LC-MS/MS analysis." *Molecules* 23, no. 8 (2018): 1985.

Guo, Long, Li Duan, Ke Liu, E-Hu Liu, and Ping Li. "Chemical comparison of *Tripterygium wilfordii* and *Tripterygium hypoglaucum* based on quantitative analysis and chemometrics methods." *Journal of Pharmaceutical and Biomedical Analysis* 95 (2014): 220–228.

Hamrapurkar, P. D., Kavita Jadhav, and Sandip Zine. "Quantitative estimation of piperine in *Piper nigrum* and Piper longum using high performance thin layer chromatography." *Journal of Applied Pharmaceutical Science* 1, no. 3 (2011): 117–120.

Hao, Chao-Yun, Rui Fan, Xiao-Wei Qin, Li-Song Hu, Le-He Tan, Fei Xu, and Bao-Duo Wu. "Characterization of volatile compounds in ten *Piper* species cultivated in Hainan Island, South China." *International Journal of Food Properties* 21, no. 1 (2018): 633–644.

Hazra, Alok K., Banti Chakraborty, Achintya Mitra, and Tapas Kumar Sur. "A rapid HPTLC method to estimate piperine in Ayurvedic formulations containing plant ingredients of Piperaceae family." *Journal of Ayurveda and Integrative Medicine* 10, no. 4 (2018): 248–254.

Heinrich, Michael, Joanne Barnes, Simon Gibbons, and Elizabeth M. Williamson. "Fundamentals of Pharmacognosy and Phytotherapy". *E-Book. Elsevier Health Sciences 2nd Edition*, (2012): 336.

Hooker, J. D. *The Flora of British India* (Vol. 5, p. 922). London: L. Reeve & Co., 1886.

Hossain, Mohammad B., Dilip K. Rai, Nigel P. Brunton, Ana B. Martin-Diana, and Catherine Barry-Ryan. "Characterization of phenolic composition in Lamiaceae spices by LC-ESI-MS/MS." *Journal of Agricultural and Food Chemistry* 58, no. 19 (2010): 10576–10581.

Iqbal, Ghazala, Anila Iqbal, Aamra Mahboob, Syeda M Farhat, and Touqeer Ahmed. "Memory enhancing effect of black pepper in the AlCl3 induced neurotoxicity mouse model is mediated through its active component chavicine." *Current Pharmaceutical Biotechnology* 17, no. 11 (2016): 962–973.

Iwamoto, Leilane Hespporte, Débora Barbosa Vendramini-Costa, Paula Araújo Monteiro, Ana Lúcia Tasca Gois Ruiz, Ilza Maria de Oliveira Sousa, MaryAnn Foglio, João Ernesto de Carvalho, and Rodney Alexandre Ferreira Rodrigues. "Anticancer and anti-inflammatory activities of a standardized dichloromethane extract from *Piper umbellatum* L. leaves." *Evidence-based Complementary and Alternative Medicine*, Article ID 9487372015: 1–8.

Jeyaprakash, Karnan, Yanung Jamoh Lego, Tamin Payum, Suriliandi Rathinavel, and Kaliyamoorthy Jayakumar. "Diversity of medicinal plants used by adi community in and around area of D' Ering wildlife sanctuary, Arunachal Pradesh, India." *World Scientific News* 65 (2017): 135–159.

Jose, M. "Indigenous aromatic and spice plants described in Van Rheed's Hortus Indici Malabarici." *Indian Journal of Applied Research* 3, no. 11 (2013): 30–33.

Kamal, Y. T., Mhaveer Singh, Shahana Salam, and Sayeed Ahmad. "Simultaneous quantification of piperlongumine and piperine in traditional polyherbal formulation using validated HPLC method." *Acta Chromatographica* 28, no. 4 (2016): 489–500.

Kaou, Ali Mohamed, Valérie Mahiou-Leddet, Sébastien Hutter, Sidi Aïnouddine, Said Hassani, Ibrahim Yahaya, Nadine Azas, and Evelyne Ollivier. "Antimalarial activity of crude extracts from nine African medicinal plants." *Journal of Ethnopharmacology* 116, no. 1 (2008): 74–83.

Karsha, Pavithra Vani, and O. Bhagya Lakshmi. "Antibacterial activity of black pepper (*Piper nigrum* Linn) with special reference to its mode of action on bacteria." *Indian Journal of Natural Products and Resources* 1, no. 2 (2010): 213–215.

Khajuria, A., N. Thusu, and U. Zutshi. "Piperine modulates permeability characteristics of intestine by inducing alterations in membrane dynamics: influence on brush border membrane fluidity, ultrastructure and enzyme kinetics." *Phytomedicine* 9, no. 3 (2002): 224–231.

Khajuria, Anu, Neelima Thusu, Ushu Zutshi, and K. L. Bedi. "Piperine modulation of carcinogen induced oxidative stress in intestinal mucosa." *Molecular and Cellular Biochemistry* 189, no. 1–2 (1998): 113–118.

Khongsai, M., H. Kayang, and S. P. Saikia. "Ethnomedicinal plants used by different tribes of Arunachal Pradesh." *Indian Journal of Traditional Knowledge* 10, no. 3 (2011): 541–546.

Khushbu, Chauhan, Solanki Roshni, Patel Anar, Macwan Carol, and Patel Mayuree. "Phytochemical and therapeutic potential of *Piper longum* Linn a review." *International Journal of Research in Ayurveda and Pharmacy* 2, no. 1 (2011): 157–161.

Kikuzaki, Hiroe, Marona Kawabata, Etsuko Ishida, Yoko Akazawa, Yoko Takei, and Nobuji Nakatani. "LC-MS analysis and structural determination of new amides from Javanese long pepper (*Piper retrofractum*)." *Bioscience, Biotechnology, and Biochemistry* 57, no. 8 (1993): 1329–1333.

Kim, Hye Jin, Wan Sook Baek, and Young Pyo Jang. "Identification of ambiguous cubeb fruit by DART-MS-based fingerprinting combined with principal component analysis." *Food Chemistry* 129, no. 3 (2011): 1305–1310.

Kim, Hye Jin, and Young Pyo Jang. "Direct analysis of curcumin in turmeric by DART MS." *Phytochemical Analysis* 20, no. 5 (2009): 372–377.

Kirtiker, K. R., and B. D. Basu. *Indian Medicinal Plants* Dehradun 3, 2nd ed. In: Kirtikar KR, Basu BD (eds). Dehra Dun, India: International Book Distributors, (1987): 2061–2062.

Kokate, C. K., H. P. Tipnis, and L. X. Gonsalves. "Anti-insect and juvenile hormone mimicking activities of essential oils of *Adhatoda vasica*, *Piper longum* and *Cyperus rotundus*." *Asian Symposium on Medicinal Plants and Spices*, Bangkok (Thailand), 15–19 September 1980.

Koul, Indu Bala, and Aruna Kapil. "Evaluation of the liver protective potential of piperine, an active principle of black and long peppers." *Planta Medica* 59, no. 5 (1993): 413–417.

Kozukue, Nobuyuki, Mal-Sun Park, Suk-Hyun Choi, Seung-Un Lee, Mayumi Ohnishi-Kameyama, Carol E. Levin, and Mendel Friedman. "Kinetics of light-induced cis-trans isomerization of four piperines and their levels in ground black peppers as determined by HPLC and LC/MS." *Journal of Agricultural and Food Chemistry* 55, no. 17 (2007): 7131–7139.

Krutika, J., V. J. Shukla, K. Nishteswar, Mandip Goyale, and Shingadiya Rahul. "Physicochemical and hptlc analysis of pippalimula (root of Piper longum LINN.)." *Indian Journal of Pharmaceutical and Biological Research* 4, no. 1 (2016): 1–6.

Kulshresta, V. K., N. Singh, R. K. Shrivastava, R. P. Kohli, and S. K. Rastogi. "A study of central stimulant activity of *Piper longum*." *Journal of Research in Indian Medicine* 6, no. 1 (1971): 17–19.

Kulshresta, V. K., R. K. Srivastava, N. Singh, and R. P. Kohli. "A study of central stimulant effect of *Piper longum*." *Indian Journal of Pharmacology* 1, no. 2 (1969): 8–10.

Kumari, Mamta, B. K. Ashok, B. Ravishankar, Tarulata N. Pandya, and Rabinarayan Acharya. "Anti-inflammatory activity of two varieties of Pippali (*Piper longum* Linn.)." *Ayu* 33, no. 2 (2012): 307.

Lakshmi, V., R. Kumar, S. K. Agarwal, and J. D. Dhar. "Antifertility activity of Piper longum Linn.in female rats." *Natural Product Research* 20, no. 3 (2006): 235–239.

Li, Song, Che Wang, Minwei Wang, Wei Li, Kinzo Matsumoto, and Yiyuan Tang. "Antidepressant like effects of piperine in chronic mild stress treated mice and its possible mechanisms." *Life Sciences* 80, no. 15 (2007): 1373–1381.

Lee, Seon A., Ji Sang Hwang, Xiang Hua Han, Chul Lee, Min Hee Lee, Sang Gil Choe, Seong Su Hong, Dongho Lee, Myung Koo Lee, and Bang Yeon Hwang. "Methylpiperate derivatives from *Piper longum* and their inhibition of monoamine oxidase." *Archives of Pharmacal Research* 31, no. 6 (2008): 679.

Lee, Seon A., Seong Su Hong, Xiang Hua Han, Ji Sang Hwang, Gab Jin Oh, Kyong Soon Lee, Myung Koo Lee, Bang Yeon Hwang, and Jai Seup Ro. "Piperine from the fruits of *Piper longum* with inhibitory effect on monoamine oxidase and antidepressant-like activity." *Chemical and Pharmaceutical Bulletin* 53, no. 7 (2005): 832–835.

Lee, Sung-Eun, Byeoung-Soo Park, Moo-Key Kim, Won-Sik Choi, Heung-Tae Kim, Kwang-Yun Cho, Sang-Guei Lee, and Hoi-Seon Lee. "Fungicidal activity of pipernonaline, a piperidine alkaloid derived from long pepper, *Piper longum* L., against phytopathogenic fungi." *Crop Protection* 20, no. 6 (2001): 523–528.

Lesiak, Ashton D., and Rabi A. Musah. "More than just heat: ambient ionization mass spectrometry for determination of the species of origin of processed commercial products-application to psychoactive pepper supplements." *Analytical Methods* 8, no. 7 (2016): 1646–1658.

Li, Ming, Xin Zhou, Yang Zhao, Dao-Ping Wang, and Xiao-Na Hu. "Quality assessment of Curcuma longa L. by gas chromatography-mass spectrometry fingerprint, principle components analysis and hierarchical clustering analysis." *Bulletin of the Korean Chemical Society* 30, no. 10 (2009): 2287–2293.

Li, Song, Che Wang, Minwei Wang, Wei Li, Kinzo Matsumoto, and Yiyuan Tang. "Antidepressant like effects of piperine in chronic mild stress treated mice and its possible mechanisms." *Life Sciences* 80, no. 15 (2007): 1373–1381.

Liu, Ling, Guoxin Song, and Yaoming Hu. "GC-MS analysis of the essential oils of *Piper nigrum* L. and Piper longum L." *Chromatographia* 66, no. 9–10 (2007): 785–787.

Lokhande, P. D., K. R. Gawai, K. M. Kodam, B. S. Kuchekar, A. R. Chabukswar, and S. C. Jagdale. "Antibacterial activity of extracts of *Piper longum.*" *Journal of Pharmacology and Toxicology* 2, no. 6 (2007): 574–579.

Majumdar, A., C. P. Shukla, R. S. Josh, and V. N. Pandey. "Hand book of domestic medicine and common ayurvedic remedies." Central Council for Research in Indian Medicine and Homoeopathy, Ministry of Health and Family Welfare, Govt of India, New Delhi, 1978.

Manavalan, R., and J. Singh. "Chemical and some pharmacological studies on leaves of Piper longum Linn." *Indian Journal of Pharmaceutical Sciences* 41, (1979): 190–191.

Manoharan, S., S. Balakrishnan, V. P. Menon, L. M. Alias, and A. R. Reena. "Chemopreventive efficacy of curcumin and piperine during 7, 12-dimethylbenz (a) anthracene-induced hamster buccal pouch carcinogenesis." *Singapore Medical Journal* 50, no. 2 (2009): 139.

Manoharan, Shanmugam, Simon Silvan, Krishnamoorthi Vasudevan, and Subramanian Balakrishnan. "Antihyperglycemic and antilipidperoxidative effects of *Piper longum* (Linn.) dried fruits in alloxan induced diabetic rats." *Journal of Biological Sciences* 6, no. 1 (2007): 161–168.

Manoj, P., E. V. Soniya, N. S. Banerjee, and P. Ravichandran. "Recent studies on well-known spice, *Piper longum* Linn." *Natural Product Radiance* 3, no. 4 (2004): 222–227.

Martins, A. P., L. Salgueiro, R. Vila, F. Tomi, S. Canigueral, J. Casanova, A. Proença Da Cunha, and T. Adzet. "Essential oils from four *Piper* species." *Phytochemistry* 49, no. 7 (1998): 2019–2023.

Mgbeahuruike, Eunice Ego, Teijo Yrjönen, H. Vuorela, and Y. Holm. "Bioactive compounds from medicinal plants: Focus on *Piper* species." *South African Journal of Botany* 112 (2017): 54–69.

Mishra, P. "Isolation, spectroscopic characterization and computational modelling of chemical constituents of *Piper longum* natural product." *International Journal of Pharmaceutical Sciences Review and Research* 2, no. 2 (2010): 78–86.

Morikawa, Toshio, Hisashi Matsuda, Itadaki Yamaguchi, Yutana Pongpiriyadacha, and Masayuki Yoshikawa. "New amides and gastroprotective constituents from the fruit of *Piper chaba.*" *Planta Medica* 70, no. 2 (2004): 152–159.

Mrutoiu, Constantin, Ioan Gogoasa, Ioan Oprean, Olivia-Florena Mrutoiu, Maria-Ioana Moise, Cristian Tigae, and Maria Rada. "Separation and identification of piperine and chavicine in black pepper by TLC and GC-MS." *JPC-Journal of Planar Chromatography-Modern TLC* 19, no. 109 (2006): 250–252.

Mujumdar, Arvind Manohar, Jayant Nilkanth Dhuley, Vinaykumar Keshav Deshmukh, Palghat Hariharan Raman, and Suresh Ramnath Naik. "Anti-inflammatory activity of piperine." *Japanese Journal of Medical Science and Biology* 43, no. 3 (1990): 95–100.

Munshi, S. R., and I. Ljungkvist. "Antifertility activity of an indigenous plant preparation (ROC101). Part 3. Effect on ultrastructure of the rat uterine luminal epithelium." *Indian Journal of Medical Research* 60, no. 12 (1972): 1791–1793.

Munshi, Safia R., Tarala V. Purandare, T. Ratnavally, and Shanta S. Rao. "Antifertility activity of an indigenous plant preparation (ROC-101). Part 2. Effect on the male reproductive system." *Indian Journal of Medical Research* 60 (1972): 1213–1219.

Murtem, G., and Pradeep Chaudhry. "An ethnobotanical study of medicinal plants used by the tribes in upper Subansiri district of Arunachal Pradesh, India." *American Journal of Ethnomedicine* 3, no. 3 (2016): 35–49.

Nabi, Shaik Abdul, Ramesh Babu Kasetti, Swapna Sirasanagandla, Thandaiah Krishna Tilak, Malaka Venkateshwarulu Jyothi Kumar, and Chippada Appa Rao. "Antidiabetic and antihyperlipidemic activity of *Piper longum* root aqueous extract in STZ induced diabetic rats." *BMC Complementary and Alternative Medicine* 13, no. 1 (2013): 37.

Naika, Raja, K. P. Prasanna, and P. S. Sujan Ganapathy. "Antibacterial activity of piperlongumine an alkaloid isolated from methanolic root extract of *Piper Longum* L." *Pharmacophore* 1, no. 2 (2010): 141–148.

Natarajan, Kavithalakshmi S., Madhusudhanan Narasimhan, K. Radha Shanmugasundaram, and E. R. B. Shanmugasundaram. "Antioxidant activity of a salt-spice-herbal mixture against free radical induction." *Journal of Ethnopharmacology* 105, no. 1–2 (2006): 76–83.

Nordén, Bo, Per Broberg, Claes Lindberg, and Amelie Plymoth. "Analysis and understanding of high dimensionality data by means of multivariate data analysis." *Chemistry & Biodiversity* 2, no. 11 (2005): 1487–1494.

Okwute, S. K. and H. O. Edharevba. "Piperine-type amides: review of the chemical and biological characteristics." *International Journal of Chemistry* 5, no. 3 (2013): 99–122.

Parganiha, R., S. Verma, S. Chandrakar, S. Pal, H. A. Sawarkar, and P. Kashyap. "In vitro anti-asthmatic activity of fruit extract of *Piper nigrum* (Piperaceae)." *International Journal of Herbal Drug Research* 1 (2011): 15–18.

Park, Byeoung-Soo, Sung-Eun Lee, Won-Sik Choi, Chang-Yoon Jeong, Cheol Song, and Kwang-Yun Cho. "Insecticidal and acaricidal activity of pipernonaline and piperoctadecalidine derived from dried fruits of *Piper longum* L." *Crop Protection* 21, no. 3 (2002): 249–251.

Parmar, Virinder S., Subhash C. Jain, Kirpal S. Bisht, Rajni Jain, Poonam Taneja, Amitabh Jha, Om D. Tyagi, et al. "Phytochemistry of the genus *Piper.*" *Phytochemistry* 46, no. 4 (1997): 597–673.

Parmar, Virinder S., Subhash C. Jain, Sangita Gupta, Sangeeta Talwar, Vivek K. Rajwanshi, Rajesh Kumar, Abul Azim, et al. "Polyphenols and alkaloids from *Piper* species." *Phytochemistry* 49, no. 4 (1998): 1069–1078.

Pattanaik, Smita, Debasish Hota, Sudesh Prabhakar, Parampreet Kharbanda, and Promila Pandhi. "Effect of piperine on the steady state pharmacokinetics of phenytoin in patients with epilepsy." *Phytotherapy Research: An International Journal Devoted to Pharmacological and Toxicological Evaluation of Natural Product Derivatives* 20, no. 8 (2006): 683–686.

Pradeep, C. R., and G. Kuttan. "Effect of piperine on the inhibition of lung metastasis induced B16F-10 melanoma cells in mice." *Clinical & Experimental Metastasis* 19, no. 8 (2002): 703–708.

Purseglove, J. W., E. G. Brown, C. L. Green, and S. R. J. Robbins. *Spices* (Vol. 1, p. 439). London: Longman, Cambridge University Press, 1981.

Rajopadhye, Anagha, Anuradha Upadhye, and Arvind Mujumdar. "HPTLC method for analysis of piperine in fruits of *Piper* species." *JPC-Journal of Planar Chromatography-Modern TLC* 24, no. 1 (2011): 57–59.

Rajopadhye, Anagha A., Tejas P. Namjoshi, and Anuradha S. Upadhye. "Rapid validated HPTLC method for estimation of piperine and piperlongumine in root of *Piper longum* extract and its commercial formulation." *Revista Brasileira de Farmacognosia* 22, no. 6 (2012): 1355–1361.

Rameshkumar, K. B., AP Anu Aravind, and P. J. Mathew. "Comparative phytochemical evaluation and antioxidant assay of *Piper longum* L. and *Piper chaba* Hunter used in Indian traditional systems of medicine." *Journal of Herbs, Spices & Medicinal Plants* 17, no. 4 (2011): 351–360.

Rao, Vidadala Rama Subba, Sagi Satyanarayana Raju, Vanka Umamaheswara Sarma, Fouriner Sabine, Kothapalli Hari Babu, Katragadda Suresh Babu, and Janaswamy Madhusudana Rao. "Simultaneous determination of bioactive compounds in *Piper nigrum* L. and a species comparison study using HPLC-PDA." *Natural Product Research* 25, no. 13 (2011): 1288–1294.

Rather, Rafiq A., and Madhulika Bhagat. "Cancer chemoprevention and piperine: molecular mechanisms and therapeutic opportunities." *Frontiers in Cell and Developmental Biology* 6 (2018): 10.

Ravindran, P. N. *Black Pepper, Piper nigrum*. Amsterdam, Netherlands: Harwood Academic, 2000.

Ravindran, P. N., R. Balakrishnan, K. Nirmal babu. "Morphometric studies on *Piper nigrum*." *JOSAC* 6, no. 1 (1997): 21–29.

Ravindran, P. N., K. Nirmal Babu. "Numerical taxonomy of South Indian Piper L. II Principal component analysis of the major taxa." *Rheedea* 6 (1996): 75–86.

Ravindran, P. N., R. Balakrishnan, K. Nirmalbabu. "Numerical taxonomy of South Indian Piper L. (*Piperaceae*) cluster analysis." *Rheedea* 2, (1992): 55–61.

Ravindran, P. N., R. Balakrishnan, K. Nirmal Babu. "Morphogenetic studies on black pepper I. Cluster analysis of black pepper cultivars." *J Spices and Aromatic Crops* 6, (1997): 9–20.

Ravindran, P. N., R. Balakrishnan, K. Nirmal Babu. "Morphometric studies on black pepper II. Principal component analysis of black pepper cultivars." *J Spices and Aromatic Crops* 6, (1997): 21–29.

Rawal, R. S., I. D. Bhatt, and Srinagar Garhwal. "Chromatographic and spectral fingerprinting standardization of traditional medicines: an overview as modern tools." *Research Journal of Phytochemistry* 4, no. 4 (2010): 234–241.

Royle, J. F. *Illustrations on Botany and Other Branches of Natural History of the Himalayan Mountains and of the Flora of Cashmere* (p. 107). London: Wm. H. Allen, 1893.

Sajem, Albert L., and Kuldip Gosai. "Traditional use of medicinal plants by the Jaintia tribes in North Cachar Hills district of Assam, northeast India." *Journal of Ethnobiology and Ethnomedicine* 2, no. 1 (2006): 33.

Salehi, Bahare, Zainul Amiruddin Zakaria, Rabin Gyawali, Salam A. Ibrahim, Jovana Rajkovic, Zabta Khan Shinwari, Tariq Khan, et al. "*Piper* species: a comprehensive review on their phytochemistry, biological activities and applications." *Molecules* 24, no. 7 (2019): 1364.

Sawangjaroen, Nongyao, Kitja Sawangjaroen, and Pathana Poonpanang. "Effects of *Piper longum* fruit, *Piper sarmentosum* root and *Quercus infectoria* nut gall on caecal amoebiasis in mice." *Journal of Ethnopharmacology* 91, no. 2–3 (2004): 357–360.

Scott, I. M., N. Gagnon, L. Lesage, B. J. R. Philogene, and J. T. Arnason. "Efficacy of botanical insecticides from *Piper* species (Piperaceae) extracts for control of European chafer (Coleoptera: Scarabaeidae)." *Journal of Economic Entomology* 98, no. 3 (2005a): 845–855.

Scott, Ian M., Evaloni Puniani, Helen Jensen, John F. Livesey, Luis Poveda, Pablo Sánchez-Vindas, Tony Durst, and John T. Arnason. "Analysis of Piperaceae germplasm by HPLC and LCMS: a method for isolating and identifying unsaturated amides from *Piper* spp extracts." *Journal of Agricultural and Food Chemistry* 53, no. 6 (2005b): 1907–1913.

Scott, Ian M., Helen R. Jensen, Bernard J. R. Philogène, and John T. Arnason. "A review of Piper spp. (Piperaceae) phytochemistry, insecticidal activity and mode of action." *Phytochemistry Reviews* 7, no. 1 (2008): 65.

Selvendiran, K., and D. Sakthisekaran. "Chemopreventive effect of piperine on modulating lipid peroxidation and membrane bound enzymes in benzo (a) pyrene induced lung carcinogenesis." *Biomedicine & Pharmacotherapy* 58, no. 4 (2004): 264–267.

Sengupta, S. "The chemistry of *Piper* species: a review." *Fitoterapia* 3 (1987): 147–166.

Shenoy, P. A., S. S. Nipate, J. M. Sonpetkar, N. C. Salvi, A. B. Waghmare, and P. D. Chaudhari. "Anti-snake venom activities of ethanolic extract of fruits of *Piper longum* L. (Piperaceae) against Russell's viper venom: characterization of piperine as active principle." *Journal of Ethnopharmacology* 147, no. 2 (2013): 373–382.

Shingadiya, Rahul, Kinnari Dhruve, V. J. Shukla, and P. K. Prajapati. "Standard manufacturing procedure and quality parameters of Kanakbindvarishta." *International Journal of Herbal Medicine* 3, no. 1 (2015): 33–36.

Shoji, Noboru, Akemi Umeyama, Nobuaki Saito, Tsunematsu Takemoto, Akiko Kajiwara, and Yasushi Ohizumi. "Dehydropipernonaline, an amide possessing coronary vasodilating activity, isolated from *Piper iongum* L." *Journal of Pharmaceutical Sciences* 75, no. 12 (1986): 1188–1189.

Singh, Ajeet, and Navneet. "Critical review on various ethonomedicinal and pharmacological aspects of *Piper longum* Linn. (Long pepper or pippali)." *International Journal of Pharmaceutical Sciences and Research* 6, no. 1 (2018): 48–60.

Singh, Amritpal, and Sanjiv Duggal. "Piperine-review of advances in pharmacology." *International Journal of Pharmaceutical Sciences and Nanotechnology* 2, no. 3 (2009): 615–620.

Singh, M., C. Varshneya, R. S. Telang, and A. K. Srivastava. "Alteration of pharmacokinetics of oxytetracycline following oral administration of *Piper longum* in hens." *Journal of Veterinary Science* 6, no. 3 (2005).

Sruthi, D., and T. J. Zachariah. "Phenolic profiling of *Piper* species by liquid chromatography-mass spectrometry." *Journal of Spices & Aromatic Crops* 25, no. 2 (2016): 123–132.

Stöhr, Jochen R., Pei-Gen Xiao, and Rudolf Bauer. "Constituents of Chinese *Piper* species and their inhibitory activity on prostaglandin and leukotriene biosynthesis in vitro." *Journal of Ethnopharmacology* 75, no. 2–3 (2001): 133–139

Sun, Cuirong, Saifeng Pei, Yuanjiang Pan, and Zhiquan Shen. "Rapid structural determination of amides in Piper longum by high performance liquid chromatography combined with ion trap mass spectrometry." *Rapid Communications in Mass Spectrometry: An International Journal Devoted to the Rapid Dissemination of Up to the Minute Research in Mass Spectrometry* 21, no. 9 (2007): 1497–1503.

Sunila, E. S., and G. Kuttan. "Immunomodulatory and antitumor activity of *Piper longum* Linn and piperine." *Journal of Ethnopharmacology* 90, no. 2–3 (2004): 339–346.

Tamuly, Chandan, Moushumi Hazarika, Jayanta Bora, Manobjyoti Bordoloi, Manas P. Boruah, and P. R. Gajurel. "In vitro study on antioxidant activity and phenolic content of three *Piper* species from North East India." *Journal of Food Science and Technology* 52, no. 1 (2015): 117–128.

Taqvi, Syed Intasar Husain, Abdul Jabbar Shah, and Anwarul Hassan Gilani. "Blood pressure lowering and vasomodulator effects of piperine." *Journal of Cardiovascular Pharmacology* 52, no. 5 (2008): 452–458.

Taufiq-Ur-Rahman, Md, Jamil Ahmad Shilpi, Muniruddin Ahmed, and Chowdhury Faiz Hossain. "Preliminary pharmacological studies on Piper chaba stem bark." *Journal of Ethnopharmacology* 99, no. 2 (2005): 203–209.

TPL. (2019). The Plant List, a working list of all plant species. http://www.theplantlist. org/1.1/browse/A/Piperaceae/Piper/ (Accessed 16 April 2019).

Tsai, Ian Lih, Fan Pin Lee, Chin-Chung Wu, Chang-Yih Duh, Tsutomu Ishikawa, Jih-Jung Chen, Yu-Chang Chen, Hiroko Seki, and Ih-Sheng Chen. "New cytotoxic cyclobutanoid amides, a new furanoid lignan and anti-platelet aggregation constituents from *Piper arborescens*." *Planta Medica* 71, no. 6 (2005): 535–542.

Tsukamoto, Sachiko, Bae-Cheon Cha, and Tomihisa Ohta. "Dipiperamides A, B, and C: bisalkaloids from the white pepper *Piper nigrum* inhibiting CYP3A4 activity." *Tetrahedron* 58, no. 9 (2002): 1667–1671.

Vaghasiya, Y., R. Nair, and S. Chanda. "Investigation of some *Piper* species for anti-bacterial and anti-inflammatory property." *International Journal of Pharmacology* 3, no. 5 (2007): 400–405.

Variyar, P. S., and C. Bandyopadhyay. "Estimation of phenolic compounds in green pepper berries (*Piper nigrum* L.) by high-performance liquid chromatography." *Chromatographia* 39, no. 11–12 (1994): 743–746.

Vedhanayaki, G., Geetha V. Shastri, and Alice Kuruvilla. "Analgesic activity of *Piper longum* Linn. root." *Indian Journal of Experimental Biology* 41, no. 6 (2003): 649–651.

Wang, H., Y. Li, W. Su, and E. Wu. "Effect of methyl piperate on rat serum cholesterol level and its mechanism of action." *Zhonggacayano* 24, no. 1 (1993): 27–29.

Wood, A. B., Maureen L. Barrow, and D. J. James. "Piperine determination in pepper (*Piper nigrum* L) and its oleoresins-a reversed phase high performance liquid chromatographic method." *Flavour and Fragrance Journal* 3, no. 2 (1988): 55–64.

Wu, E., and Z. Bao. "Effects of unsaponificable matter of *Piper longum* oil on cholesterol biosynthesis in experimental hypocholestrolaemic mice." *Honggacayano* 23, no. 4 (1992): 197–200.

Wu, Shihua, Cuirong Sun, Saifeng Pei, Yanbin Lu, and Yuanjiang Pan. "Preparative isolation and purification of amides from the fruits of *Piper longum* L. by upright counter-current chromatography and reversed-phase liquid chromatography." *Journal of Chromatography A* 1040, no. 2 (2004): 193–204.

Xia, Yuanyuan, Guangli Wei, Duanyun Si, and Changxiao Liu. "Quantitation of urso-lic acid in human plasma by ultra-performance liquid chromatography tandem mass spectrometry and its pharmacokinetic study." *Journal of Chromatography B* 879, no. 2 (2011): 219–224.

Yadav, Mukesh Kumar, June Choi, and Jae-Jun Song. "Protective effect of hexane and ethanol extract of Piper longum L. On gentamicin-induced hair cell loss in neonatal cultures." *Clinical and Experimental Otorhinolaryngology* 7, no. 1 (2014): 13.

Yang, K. Yang, Kuo-Yi, Lei-Chwen Lin, Ting-Yu Tseng, Shau-Chun Wang, and Tung-Hu Tsai. "Oral bioavailability of curcumin in rat and the herbal analysis from *Curcuma longa* by LC-MS/MS." *Journal of Chromatography B* 853, no. 1–2 (2007): 183–189.

Yang, Young-Cheol, Sang-Guei Lee, Hee-Kwon Lee, Moo-Key Kim, Sang-Hyun Lee, and Hoi-Seon Lee. "A piperidine amide extracted from *Piper longum* L. fruit shows activity against Aedes aegypti mosquito larvae." *Journal of Agricultural and Food Chemistry* 50, no. 13 (2002): 3765–3767.

Yende, Subhash R., Vrushali D. Sannapuri, Niraj S. Vyawahare, and Uday N. Harle. "Antirheumatoid activity of aqueous extract of *Piper longum* on freunds adjuvant-induced arthritis in rats." *International Journal of Pharmaceutical Sciences and Research* 1, no. 9 (2010): 129–133.

Zachariah, T. John, A. L. Safeer, K. Jayarajan, N. K. Leela, T. M. Vipin, K. V. Saji, K. N. Shiva, V. A. Parthasarathy, and K. P. Mammootty. "Correlation of metabolites in the leaf and berries of selected black pepper varieties." *Scientia Horticulturae* 123, no. 3 (2010): 418–422.

Zaveri, Maitreyi, Amit Khandhar, Samir Patel, and Archita Patel. "Chemistry and pharmacology of *Piper longum* L." *International Journal of Pharmaceutical Sciences Review and Research* 5, no. 1 (2010): 67–76.

Zhao, Yang, Li Wang, Yuanwu Bao, and Chuan Li. "A sensitive method for the detection and quantification of ginkgo flavonols from plasma." *Rapid Communications in Mass Spectrometry: An International Journal Devoted to the Rapid Dissemination of Up to the Minute Research in Mass Spectrometry* 21, no. 6 (2007): 971–981.

Index

Printed in the United States
by Baker & Taylor Publisher Services